HISTORY OF

SCIENCE

The story of science is possibly the strangest tale ever told in human history. We don't know quite where it began. We have no idea how, when, or if it will end. What we do know is that humanity's relentless need to question and explore has propelled us from hunter-gatherers with a tenuous grasp on fire to a world-straddling civilisation that can split the atom, transmit information through the air, and even step outside the boundaries of our own Earth and into space beyond. No history of science can include all of the incredible events, people, theories and discoveries that have revolutionised our understanding of our place in the universe. In these pages we've concentrated on some of the biggest and most disruptive events that have transformed us and the world around us, but there is so much more to discover...

CONTENTS

6 **TIMELINE OF SCIENCE**
Some of the key discoveries in the history of science

10 **BEFORE THE BRONZE AGE**
How the Chalcolithic set the stage for humanity's early leap into metalwork

14 **HOW BRONZE BEGAN**
Discover this key turning point in human technology

20 **THE BUILDING BLOCKS OF SCIENCE**
How the Iron Age gave rise to new ways of thinking

28 **THE FIRST ASTRONOMERS**
What the ancient Babylonians saw

30 **THE CRUCIBLE OF ALCHEMY**
How a unique Egyptian city nurtured early proto-science

36 **THE ISLAMIC GOLDEN AGE**
Explore the advanced technologies of the medieval Middle East

40 **ALCHEMY ENTERS EUROPE**
How translators brought alchemy north and west

44 **THROUGH A NEW LENS**
Why the development of optics was so important to science

46 **MEDICINE IN THE MEDIEVAL PERIOD**
The bizarre remedies and researches of medieval healers

50 **NATURAL PHILOSOPHY**
The philosophical forerunner to both biology and physics

54 **NICOLAUS COPERNICUS**
How a stargazing monk set a revolution in motion

56 **THE THIRST FOR SCIENTIA**
The polymaths of the Renaissance

60 **TYCHO BRAHE**
Meet the man who first calculated planetary motion

62 **GALILEO GALILEI**
The father of modern astronomy and his controversial career

64 **JOHANNES KEPLER**
The man who defined the three laws of planetary motion

66 **SIR ISAAC NEWTON**
Discover the incredible ideas of the father of modern physics

68 **ANTONIE VAN LEEUWENHOEK**
The cloth merchant who looked at microscopic life through a lens

72 **WHEN THE WORLD CHANGED**
How the Scientific Revolution changed the world

76 **AN AGE OF TRANSITION**
When alchemy transmuted into chemistry

80 **CHARLES DARWIN**
How religion nearly obscured the theory of evolution

86 **EDISON VS TESLA**
The electrifying story of the Current Wars

92 **MARIE CURIE**
A pioneer in nuclear physics and chemistry

96 **NIELS BOHR**
Modelling the atom

98 **ALBERT EINSTEIN**
The iconic scientist who changed our view of everything

102 **IT BEGAN WITH A BANG**
The theory that upended our ideas about the universe

104 **UP, UP AND AWAY**
A brief look at the history of aviation and powered flight

106 **PIONEERS OF COMPUTING**
The world goes digital

110 **RISE OF THE SPACE AGE**
How the Cold War opened a new scientific frontier

116 **UNLOCKING DNA**
Discovering the secret of life and the study of genetics

118 **WORLDS WITHIN WORLDS**
Explaining the building blocks of the universe

120 **STEPHEN HAWKING**
Revealing the theories of a true titan of science

126 **DOLLY THE SHEEP**
The first mammal successfully cloned

128 **THE HIGGS MECHANISM**
The 'God Particle' goes from theory to proven fact

TIMELINE OF SCIENCE

THESE ARE SOME OF THE KEY EVENTS AND DISCOVERIES IN THE HISTORY OF SCIENCE

WRITTEN BY **EDOARDO ALBERT**

UP TO TWO MILLION YEARS AGO

FIRE

The use of fire is so old it predates us: our evolutionary ancestor, homo erectus, was using fire in a controlled manner two million years ago. By 400,000 years ago, fire had become a commonplace tool of our ancestors, being used for cooking, warming and, it seems, as a centrepiece around which people sat and talked while staring into the flames. Some things don't change. When homo sapiens replaced homo erectus, we still sat around fires, staring into the flames.

5 JULY 1687

TIED TOGETHER BY UNIVERSAL LAWS

"Nature, and Nature's Laws, lay hid in Night. God said, Let Newton be! and all was Light!" That was the epitaph composed for Isaac Newton by Alexander Pope, and it shows the impact his work had. Newton was responsible for the single most important book in the history of science: *Philosophiæ Naturalis Principia Mathematica* ('The Mathematical Principles of Natural Philosophy'). In it, he established the laws of motion and of universal gravitation as well as developing calculus as a key mathematical tool. Not only did Newton's laws describe planetary motion, but they showed that the universe itself, from the orbits of the planets to the fall of a single apple, could be described in precise mathematical terms. Not only was the universe rational, it was mathematical, and it obeyed mathematical rules.

1673

WORLDS WITHIN

In 1673, the Royal Society published a letter it had received from a Dutch draper, Antonie von Leeuwenhoek. The Royal Society, the foremost scientific institution in the world, was not in the habit of publishing letters from drapers, Dutch or otherwise, but what it read in Leeuwenhoek's letter was so important that they had little choice. This first letter described what Leeuwenhoek had seen when looking through his own magnifying lenses at moulds, lice and bees. It would be the first of nearly 200 letters that Leeuwenhoek sent to the Royal Society. In these letters, Leeuwenhoek described a new world: the world within. Bacteria, red blood cells, protozoa, sperm cells. These were just a few of his discoveries in the world of the very small. What's more, Leeuwenhoek conclusively disproved the old idea that lower forms of animal life spontaneously generated from their surroundings. The telescope and now the microscope were expanding the universe outwards and inwards to a scale previously undreamed.

C.1200BCE – C.600BCE

THE AGE OF IRON

It's a rock. It's hard, unyielding and it usually has a dull red tinge. It certainly doesn't look useful for anything other than bashing somebody on the head with. But somewhere around 3200 years ago, in Mesopotamia, the land between the rivers, somebody heaped up these stones and heated them until, wonder of wonders, they melted. To take something that was the very definition of hardness, rock itself, and turn it into a liquid was a marvel in itself. But when that liquid cooled, further wonders ensued: it made a material harder, stronger and tougher than anything people had before. So, of course, they used it to make swords and spears: the Hittite Empire was founded on their use of iron weapons. But iron could also make ploughs and axes, trains and planes and automobiles. It ranks with agriculture as the most transformative discovery in human history.

5TH TO 13TH CENTURIES CE

A RATIONAL WORLD

There's a reason that advanced cultures, such as China and the Islamic world, didn't develop science in quite the same way that Europeans did. For science to be even conceivable as an enterprise, the world has to make sense. What's more, we have to believe that it makes sense. In the Islamic world, the great scholar al-Ghazali had settled this question in the 12th century: everything happened by God's direct will. In China, Confucianism was concerned with society and politics, not nature. Both cultures were highly technically advanced, but they didn't question the 'why' of how things behaved in the way that the West did, a key difference that fuelled the conceptual development of western science. In the Christian world, thinkers such as Augustine, Bede, William of Occam and Thomas Aquinas argued that a rational God had created a rational world. What's more, by seeking to understand that rational world, light could be cast on the God that made it. This was the fundamental axiom necessary for early science to be possible: that the world makes sense and, furthermore, that we can understand it and that it is intrinsically good for us to understand it more deeply.

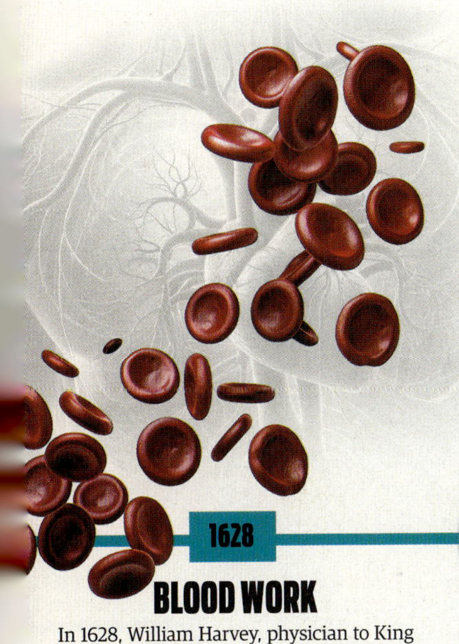

1628

BLOOD WORK

In 1628, William Harvey, physician to King Charles I, published a book, *Exercitatio Anatomica de Motu Cordis et Sanguinis in Animalibus* ('Anatomical Exercise on the Motion of the Heart and Blood in Animals') that established, for the first time, how blood circulates through the body, pumped from the heart. Previously, physicians had thought there were separate circulatory systems, the venous and arterial systems, originating from the liver and the heart and lungs respectively. Harvey conclusively showed that it was a single integrated circulatory system, flowing out from the heart and returning to it, with the one-way valves that are found in veins all orientated towards the heart and not against gravity, as had been thought. This discovery, overturning Galen's ancient theory, set medicine firmly on the path of science.

30 NOVEMBER 1609

WORLDS WITHOUT

In 1609, a young Italian named Galileo Galilei heard about the new device that could make distant objects seem much closer. Working from some garbled descriptions of the device, the young man made his own version. Looking through it, objects seemed three times bigger. But, being restless and impatient, Galileo set about improving his 'telescope', until it made objects eight times bigger. Galileo demonstrated his new telescope to the Venetian Senate on 25 August, which rewarded him with a lifetime professorship at the University of Padua and a doubling of his salary. Galileo continued working on his telescopes, increasing the magnification to 20 times. Then, at the end of November, he pointed his telescope up at the sky. Through the lens, Galileo saw that the Moon's surface was shadowed and, he quickly realised, these shadows were cast by mountains on the Moon. Through the next few months he continued observing, discovering that Jupiter had moons in orbit around it and that there were a host of stars invisible to the naked eye. It was as if the lid had been taken off: the universe had suddenly become much, much bigger.

1088

THE UNIVERSITY

In 1088, an Italian law scholar began teaching students in Bologna. These young men were the first students and Bologna was the first university. The concept of universities as independent institutions where scholars could gather, teach, research and dispute, was quickly established, with other institutions being founded in Paris, Oxford, Salamanca, Cambridge and elsewhere. These new medieval institutions zealously guarded their independence, establishing the ideals of intellectual inquiry and rigorous testing of ideas that would prove foundational to the whole scientific enterprise, as well as giving future scientists places where they could live, eat and carry out their experiments in reasonably congenial conditions.

1735

A PATH THROUGH THE JUNGLE

Newton had it easy. There were only six planets, including the Earth, known when he was developing his law of gravitation. The first biologists and botanists had a much more difficult task. They were faced with a bewildering variety of creatures and no obvious way to work out what was what and how it related to other organisms. To make a start on this, they needed a system of classification, something that could accommodate the creatures and plants already known and that could be expanded to cope with the new plants and animals being discovered in the 18th century. Step forward Carl Linnaeus. His system of binomial naming is the basis for all modern biology, accommodating organisms as diverse as homo sapiens and stapelia gigantea (the carrion plant that, when it flowers, produces the smell of rotting flesh).

1789

THE CHEMICAL REVOLUTION

Chemistry came late to the scientific party. While physics, biology and medicine had had their foundations laid, it took until the end of the 18th century for chemists to develop a framework for their science. The man most responsible for this was Antoine Lavoisier. A Frenchman, Lavoisier combined his scientific research with a public life deeply involved with the reform of the tottering French state. The Chemical Revolution he fostered was based upon some key principles he proposed, as well as a number of discoveries that demonstrated their utility. The chief of these was the conservation of mass, the idea that in any chemical reaction the mass of the reactants is conserved. Lavoisier also demonstrated that combustion and respiration came about through reactions with oxygen. Unfortunately for Lavoisier, his work for the French state made him a marked man after the French Revolution, even though he himself saw the Revolution as a chance to reform the state. During the Terror, he was condemned and sent to the guillotine.

24 TO 29 DECEMBER 1927

THINGS GET WEIRD

For six days, in Brussels, Belgium, 28 men and one woman gathered for what was probably the most intellectually high-powered conference in history. Of the attendees, 17 were or would become Nobel laureates. Among those present were Albert Einstein, Niels Bohr, Werner Heisenberg, Paul Dirac, Max Born, Ernst Schrödinger and Marie Curie, and what they were talking about was quantum mechanics. There, they thrashed out the details of the most successful physical theory yet devised. Quantum theory describes the behaviour of things at the atomic scale and smaller where, it turns out, waves can be particles and particles waves, where position or momentum can be established but not both, and where Schrödinger's cat sits, smiling at our mystification.

23 NOVEMBER 1924

IT'S A BIG OLD UNIVERSE OUT THERE

Looking through their telescopes, astronomers had discovered objects that did not seem to be stars. They were luminous, but their light appeared more diffuse than ordinary stars, so they were named nebulae to indicate that their light was spread out. By 1920, the size of the Milky Way had been established, but most astronomers thought that was as far as the universe went. But a young astronomer, Edwin Hubble, using the state-of-the-art 100-inch telescope on Mount Wilson, California, discovered a Cepheid variable star in the Andromeda nebula and, from that, he was able to work out how far away it was. The answer established beyond doubt that Andromeda was not a nebula but a separate galaxy, far, far away. The universe had suddenly got a whole lot bigger.

JULY 1830

DEEP TIME

In July 1830, Charles Lyell, a Scotsman, published the first volume of his *Principles of Geology*. In it, he unveiled, for his unsuspecting Victorian audience, a vista of time so deep that it dizzied the mind. Rather than hundreds or thousands of years, the Earth was hundreds of thousands, possibly even millions of years old. In his book, Lyell set out to demonstrate that the processes geologists saw around them, such as erosion and deposition, were sufficient to form all the physical features of the Earth. But for these slow processes to build up mountains and carve out river valleys, a new timescale was necessary, one that stretched back into a deep, deep past. Lyell opened the well of time. Another man would shortly look down it and draw out the next key breakthrough in science.

27 DECEMBER 1831

THE TREE OF LIFE

On the above date, a young Charles Darwin went aboard the HMS Beagle on a five-year voyage charting the coast of South America. For reading on the voyage, Darwin had the first two volumes of Charles Lyell's *Principles of Geology*, which opened up to him the expanse of time over which he was travelling. During the course of the voyage, in particular from the related but different species of the Galapagos Islands, Darwin developed the germ of his idea that species evolved by natural selection. On his return, Darwin spent 20 years refining his idea and gathering evidence for it before, on 22 November 1859, he published *On the Origin of Species*. His theory would become the unifying theory for all the biological sciences, ensuring its place as a scientific milestone.

1905 AND 1915

IT'S ALL RELATIVE

The universe of Newtonian mechanics was precise and elegant, like the intricate wheels of an old-fashioned mechanical watch, each turning the next. And for over 200 years it had described everything from the motion of the stars to the rebound of the balls on a snooker table. But, first in 1905 and then again in 1915, Albert Einstein published two papers that made this world curve. His theories of special and general relativity changed the geometry of the universe from one of the three spatial dimensions set in time, to a unified space-time geometry, and in doing so, things curved. It turned out that the Earth orbits the Sun because the very structure of space-time curves and the Earth rides the curve like a surfer. The universe, it turned out, was stranger than we thought. And it was about to become stranger than we could have imagined.

31 MAY 1881

VACCINATION

On this date in 1881, a team of researchers who worked for Louis Pasteur injected 58 sheep, two goats and ten cows with a carefully cultivated preparation of the anthrax bacillus. Of these animals, half had previously been given injections of an attenuated version of the anthrax bacteria. Two days later, a crowd of over 200 people, including Pasteur himself, came to view the results. They were clear to see. Half the animals were fine, the other half, the half that had not been given the extra injections, were either dead or dying. Louis Pasteur and his team had developed vaccination. This differed from the previous smallpox vaccination, which used a naturally occurring weaker form of the disease to inoculate against the disease, by actually weakening or killing the infectious organism itself before injecting it. In terms of the greatest good, Pasteur stands head and shoulders above all the rest, for his discovery has saved the lives of hundreds of millions of people

BEFORE THE BRONZE AGE

THE TRANSITIONAL PERIOD AT THE END OF THE NEOLITHIC WAS A MYSTERIOUS AND EXCITING ERA ON THE BRINK OF A TECHNOLOGICAL AND CULTURAL REVOLUTION

WRITTEN BY **APRIL MADDEN**

Among the earliest and most disruptive events in the history of science is metalworking, which is usually thought of as beginning in the Bronze Age. But this massive innovation didn't arise in a vacuum. It can be tempting to think of the inception of metalworking as a sudden jolt that shook Stone Age society to its core, but it was in fact the next stage in a process that had begun centuries beforehand.

The anthropological three-age system is divided into a trinity of evolving technologies, starting with the flint tools of our earliest modern-human ancestors and progressing to the use of bronze and then iron. But the 'Stone Age' itself is something of a broad term, and the primitive cavemen of the popular imagination are nothing like the (sometimes surprisingly sophisticated) people who actually lived in it. For a start, there were several eras within the Stone Age, encompassing many centuries. These are the Paleolithic, Epipaleolithic, Mesolithic, and Neolithic - the period in which humans adopted agriculture. At this point, many human communities made the switch from a nomadic hunter-gatherer lifestyle to a settled agrarian one, building small settlements, cultivating crops and herding animals. From such small beginnings, it can seem like the Bronze Age is an inexplicably swift and major leap forward. But there is another period, one that straddles the gap between the mysteries of prehistory and our more solid knowledge of history. It's a period of what's known as 'proto-history' - an era in which some civilisations had writing and some didn't, and those who had recorded some vital historical details about those who hadn't. This liminal, interstitial era at the end of the Neolithic, immediately before the Bronze Age began, is known variously as the Chalcolithic or Eneolithic: the Age of Copper.

The Belovode archaeological site on Mount Rudnik in Serbia contains the earliest securely dated evidence for deliberate copper smelting. Could its copper smiths have been part of a mysterious, technologically advanced culture that swept through Europe and Asia during the Eneolithic?

Russian folklore tells of a place in the mythical Kingdom of Opona, a utopian earthly paradise at the edge of the Earth, where the inhabitants were all happy, healthy, well-fed and well-off. This mystical land was known as Belovode, which is also the name of an archaeological site on Mount Rudnik ('mine') in Serbia. Home to Illyrians, Celts, Romans and Slavs over the centuries, amid the legends of ancient castles and cursed queens that litter its beech-clad slopes, Belovode contains the earliest securely dated evidence for high-temperature copper smelting in the world, from around 5000 BCE.

Smelting is the process of heating metal-bearing ore to extract the metal within. We know that the first metals to be deliberately smelted were tin and lead, thanks to some cast lead beads found at Çatal Höyük in Turkey that date to at least 6500 BCE. This predates writing by around 3500 years, so we don't know anything about how or why this innovation arose. What we do know is that other cultures across the ancient Near East soon began experimenting with smelting too: a lead bangle found in Iraq dates to around the same time as the Turkish beads. Lead is relatively

"THIS ERA AT THE END OF THE NEOLITHIC, IMMEDIATELY BEFORE THE BRONZE AGE BEGAN, IS KNOWN VARIOUSLY AS THE CHALCOLITHIC OR ENEOLITHIC: THE AGE OF COPPER"

Artist's impression of the heavily fortified Copper Age Los Millares proto-city discovered in Andalucía, Spain. The town, which participated in iconic European visual trends Beaker culture, Megalithism, and Symbolkeramik pottery, had a large building dedicated to copper smelting among its smaller dwellings, which housed around 1,000 people

rare in Iraq, but we can't say for sure whether the material and technology were transported and traded, or if this instance was an independent discovery. A theoretical argument could be made for the beginnings of trade: the Copper Age is the period during which the wheel was invented and the horse domesticated; both innovations eventually made goods easier to transport and trade over larger distances.

Tin and lead are soft metals – they can be liquified over the heat of a hearth or campfire – which makes them easy to smelt, even accidentally, but unsuitable for many of the uses that eventually saw metal replace stone: the creation of tools and weapons. Copper, on the other hand, requires a smelting temperature that's around 200 degrees Celsius higher than an ordinary wood fire can produce. We don't know how those first copper smiths achieved the requisite heat: the theory is by using pottery kilns, which can get up to a much higher temperature. We know that there were significant Neolithic pottery cultures in the times and places where the earliest smelted copper artefacts have been found, so the kiln hypothesis is our current best guess. Pottery culture certainly affected other game-changing innovations in the Copper Age; the invention of the potter's wheel predates that of the wheeled vehicle by at least several centuries. The first known writing – the cuneiform of ancient Sumer (now Iraq) – was inscribed into tablets of clay.

The copper smiths of Eneolithic Belovode were almost certainly speakers of a variant of the Proto-Indo-European language, the common ancestor of many Old World tongues and cultures. According to the Kurgan Hypothesis, originally formulated by archaeologist Marija Gimbutas in the 1950s, the speakers of Proto-Indo-European swept into and across Europe and Asia from the Pontic-Caspian steppe (part of both Russia and southern Ukraine) from the 6th millennium BCE onwards. This theory was further refined by anthropologist David W Anthony in 2007 with the publication of his book *The Horse, the Wheel, and Language: How Bronze-Age Riders from the Eurasian Steppes Shaped the Modern World*. Anthony's theory describes how a period of pre-Bronze Age climate change caused significant movements of early nomadic herders, which led to linguistic, cultural and technological interchanges between different groups. These herders were the likeliest candidates for the earliest domestication of the horse, but the wider geographical ranges of the highly mobile nomadic groups created by horse domestication caused increased instances of warfare, creating a need for protection and more advanced weaponry. As the Copper Age became the Early Bronze, settlements began to be more heavily fortified; the Eurasian Sintashta culture invented

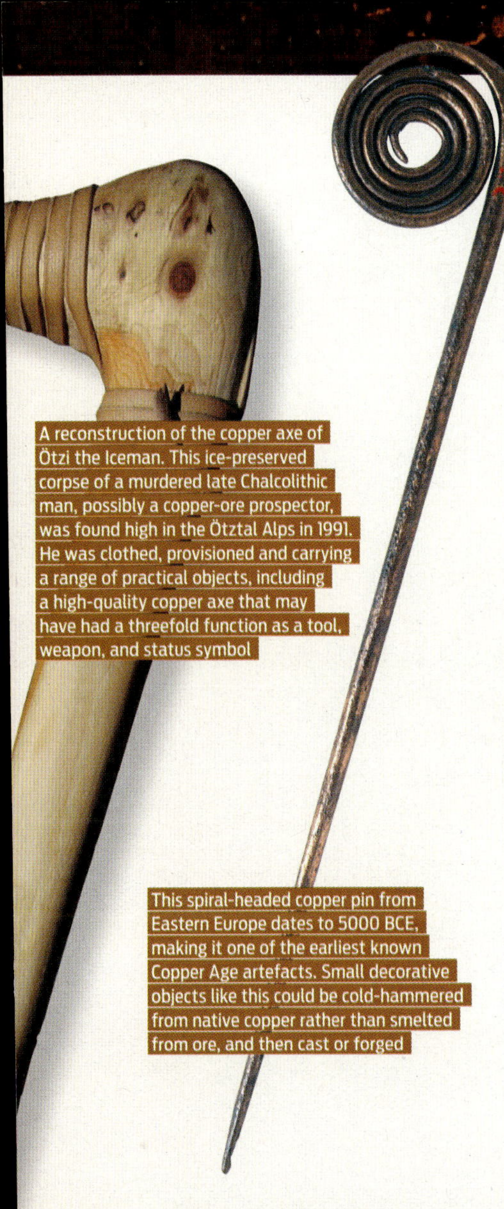

A reconstruction of the copper axe of Ötzi the Iceman. This ice-preserved corpse of a murdered late Chalcolithic man, possibly a copper-ore prospector, was found high in the Ötztal Alps in 1991. He was clothed, provisioned and carrying a range of practical objects, including a high-quality copper axe that may have had a threefold function as a tool, weapon, and status symbol

This spiral-headed copper pin from Eastern Europe dates to 5000 BCE, making it one of the earliest known Copper Age artefacts. Small decorative objects like this could be cold-hammered from native copper rather than smelted from ore, and then cast or forged

chariots, whose use rapidly spread to the Middle East and Egypt, as did the copper that the Sintashta mined and traded intensively. Anthony's theory describes what can be archaeologically evidenced by what is referred to as the Seima-Turbino Phenomenon, which describes a pattern of stylistically similar artefacts and burial rituals from a relatively small time period during the Early Bronze Age (2100 BCE to 1900 BCE) that appear in sites thousands of miles apart across vast swathes of Europe and Asia. This, too, suggests the rapid mass-migration of a single cultural group, one that had significantly advanced metallurgical techniques compared to others of the era. Whether they were Sintashta or another group remains to be seen (the widespread Beaker culture is another potential candidate), but one thing is certain: these late Chalcolithic metalworkers were on the bleeding edge of the Bronze Age, and they took it, along with their language, culture and technology, with them wherever they went, to places as far apart as Mongolia, Finland, and Iran.

But were they the only people bringing copper (and later bronze) manufacture to the world? Absolutely not. We know that metalworking originated independently in the Americas, which at that point had no known contact with the Old World. Copper artefacts from sites in South and Central America, as well as the United States, can be dated to similar timeframes to those in Europe and Asia; the earliest known in South America are those from the Andean civilisations, but in North America they date even further back, to around 5000 BCE. However, North American copper artefacts from this period are a technological exception: they weren't smelted, cast or forged. Raw copper was abundant around the Great Lakes area, whose rocks had been scoured by glaciation during the last Ice Age, bringing the metal within much closer to the surface, and was obtained by creating cracks in the rocks using a combination of fire and water – essentially a primitive sort of fracking – and pounding with a hammerstone. The copper was then cold-hammered into the desired shape. Ancient South America, however, developed a full grasp of metallurgy, with evidence for smelted copper dating to 2000 BCE in the central Andes. Although no known Andean copper artefacts date back that far, results from an ice-core chemical analysis from the Illimani glacier, published in the scientific journal *Nature* in 2017, show that central Andean cultures were smelting enough copper at this time for the man-made pollution from their activity to be recorded in the ice of the nearby glacier.

Copper is, of course, an essential ingredient in the makeup of bronze, and the smelting cultures soon discovered that molten metals could be mixed into alloys. The first bronzes, found in the Middle East, are alloys of copper and arsenic, but the tough, toxic-fume-free bronze of the Bronze Age proper is made of copper and tin. These two metals are rarely found naturally together: only three lode sites worldwide carry deposits of both – one in Britain, one in Thailand, and one in Iran. The people of the Bronze Age would have to trade far afield to obtain the materials that made their civilisations possible. Perhaps it was the clever, highly mobile people of the Copper Age that gave them the idea to do so, along with many other technologies and ideas that would change the world radically.

AFRICA: BYPASSING BRONZE

We think of the Bronze Age as a necessary step on the road to progress, but there are some areas and cultures that didn't bother with bronze and went straight from copper to iron. Studies suggest that the Nok culture of Nigeria in West Africa was producing iron from around 1000 BCE onwards. African ironwork utilised a particular kind of smelting technology known as blooming that was less popular in the rest of the Old World, which soon after the advent of the Iron Age (c 1200 BCE) began to adopt early blast furnace techniques.

The legacy of colonialism means that ancient African metallurgy is an under-studied subject – for decades, historians persistently clung to a belief that early metalworking techniques had been imported to Africa via either Egypt or Phoenician Carthage (Tunisia), and that iron had not been produced in the continent at all until the 19th century. Research into ancient African bloomeries still comes up against far more rigorous questioning than other archaeological smelting sites worldwide, with arguments raised about everything from the sources and storage methods of the charcoal that powered the bloomeries, to the physical context of the iron artefacts found, to problems inherent in accurate radiocarbon dating worldwide for the period 800 to 400 BCE. While we can now say for certain that African cultures were producing and using iron prior to the colonial period, what's still a source of debate is whether iron production arose independently in Africa, or whether it was imported from nations on the edge of the continent and the trade routes that had access to Eurasian trade and technology.

A bloomery in action. Bloomeries were the first form of smelter capable of producing high enough temperatures to smelt iron. The iron that's smelted must be further worked and wrought before it can be used to craft objects

HOW BRONZE BEGAN

THE COMING OF BRONZE WAS A TURNING POINT IN HISTORY WHEREVER IT APPEARED, AND NOTHING WOULD EVER BE THE SAME AGAIN

WRITTEN BY **BEN GAZUR**

The Copper Age saw the first smelting and casting of metal tools. The bright gleam of copper and its ability to be shaped at need would have made it one of the wonders of the age. Yet for all its brilliance there is a flaw in copper - it flashes prettily but bends and deforms easily. You would have a hard time hammering a copper nail into wood. Something harder was needed if metal was to become truly useful. With smelting already invented, all that was needed was for people to learn to mix different kinds of metals together to make alloys. The Bronze Age was to be the age of the alloy.

Inventing Bronze

It can seem counter-intuitive, but alloying two metals together that are both soft can create a material that's much harder than either individually. To understand why you have to work out exactly what is happening within a metal.

Metallic elements in their solid form have a crystalline structure. The atoms arrange themselves into set patterns that repeat throughout the material. The atoms form layers surrounded by a soup of electrons. In a pure metal these layers will be regular and often form quite a malleable material, as atoms can easily be moved around within the structure with layers slipping past their neighbours without much force.

When an alloy is made, however, a new element is introduced to the mix. The new element's atoms have different properties and sizes to the original metal's atoms and so do not fit regularly into the crystal structure. It is like mixing baseballs and footballs together. Disorder is introduced to the material and this can bring a number of benefits. Without neat layers more force is required to push layers past each other, making for a tougher material. The new structure of the alloy may also have a lower melting point than the pure metal, making casting and reforging it more simple.

There is still controversy about where the first bronze items were made. Dating metals can be difficult even when found in a sound archaeological context. Current thought is that bronze technology emerged in many locations at different times once the ability to smelt copper was in place. It does seem likely though that the first alloys were entirely accidental.

In many places copper and arsenic are found together. Egyptian copper ores can have significant amounts of arsenic in them and so when the copper from those ores was smelted, an unintentional alloy was created called arsenic bronze. This alloy produced more durable tools and when molten it flowed into moulds more readily. At some point it must have been recognised that objects cast from these sources were more useful because increasing levels of arsenic, far beyond what is possible by accidental contamination, were later added to copper. Workers in arsenic bronze had a sophisticated knowledge of their craft. If arsenic bronze is created with more then 2% arsenic then the resulting metal is brittle and easily splintered. Objects from many places and times show that the amount of arsenic was very finely controlled around this level

> "ALLOYING TWO METALS TOGETHER THAT ARE BOTH SOFT CAN CREATE A MATERIAL THAT'S MUCH HARDER THAN EITHER INDIVIDUALLY"

Early metal workers are likely to have discovered alloys by accident at first, before beginning to explore the possibilities of this new technology

Moulds chiselled from stone blocks allowed near endless replication of of bronze items like this sword from ~1900 BCE Italy

The ability of bronze to hold a sharp edge gave it a place in the tools of many crafts, such as this adze from Ban Chiang

– quite the achievement for people with no knowledge of what an element was. It may be that the garlic-smell of arsenic compounds when heated and beaten with a hammer guided bronze makers in their work.

When arsenic bronze is work-hardened, beaten with a hammer, it forms a structure that is much more able to hold a cutting edge. Arsenic bronze remained in use in Egypt from the pre-dynastic era of ~5000 BCE to the Middle Kingdom of ~2000 BCE.

While the ancient metallurgists would have been entirely unaware of the chemistry they were performing, there are other benefits to using arsenic when smelting copper. If oxygen is present when copper is smelted it can lead to the creation of copper oxide, which would reduce the strength of objects made with it. Arsenic reacts with oxygen and creates arsenic oxides that vaporise and leave the metal free from oxygen contamination. As we shall see this vapour had other, unintended, consequences.

Another reason that arsenic may have been added to copper during smelting is one of aesthetics. Increasing the amount of arsenic in the alloy changes the final appearance of the polished bronze. The more arsenic, the more silvery the final result appears.

In the Caucasus, bronze daggers have been found with arsenic-rich external layers that may have been used simply to change the appearance of the metal.

Despite tin being a far more rare material and difficult to locate and transport around 3000 BCE, in many places in the Middle East tin became the preferred element to add to copper to create bronze. Why would a common ingredient be replaced with a rare and expensive one? There are material advantages to tin bronze but the dangers of working with arsenic have to be acknowledged.

In the ancient pantheons of several religions there was a specialised god of many crafts, and smithing was no exception. For the Greeks, Hephaestus was the god metal workers prayed to, while to the Romans he was Vulcan. In both cultures he is called 'lame' and is depicted as limping about. Other smithy gods such as Wieland, Volander, and Ilmarinen are also shown as disabled in some form or other. While mythology offers an explanation for his disability, it has been suggested that the smith god's problems actually arose from the memory of metal workers lamed by exposure to arsenic in their work.

Chronic arsenic poisoning could occur if workers breathed in the arsenic-laden vapours that smelting produced. Damage to the peripheral nerves is common in those poisoned in this way, and a limp or dragging foot sometimes develops. For those with the most exposure to arsenic, terrifying symptoms like vomiting, convulsions, and death could have been the result of their pursuit of useful tools. The coming of tin to the world of bronze manufacture may have been sped by the desire not kill those most skilled in the art of making bronze.

As well as being less toxic than arsenic, tin has other advantages. When six to ten per cent tin is added to copper you get a bronze that is strong, resistant to salt water corrosion, and holds a sharp edge. Different areas used differing ratios of tin to copper, however, with Chinese bronze swords often

"IT HAS BEEN SUGGESTED THAT THE SMITH GOD'S PROBLEMS AROSE FROM THE MEMORY OF METAL WORKERS LAMED BY EXPOSURE TO ARSENIC"

having up to 20 per cent tin which gave them remarkably sharp edges but left the whole blade brittle. The differing qualities of bronze that could be produced depending on the amount of tin that was added would later be exploited. Mild bronze, made with 6 per cent tin, could be beaten into sheets and was used for helmets and armour. Classic bronze of 10 per cent tin was used in swords and tools.

Tin bronzes are as strong as arsenic bronzes and the molten metal flows into moulds smoothly to fill all available space. The real advantage is that working with tin allows for identical batches of bronze to be made each time. The danger of arsenic is in the vapours it produces. Since arsenic can be driven off in different amounts depending on a range of conditions from temperature to availability of oxygen, you can never be totally certain of how much arsenic will remain in the finished alloy. Tin does not vaporise readily. Arsenic-rich ores can also have varying amounts of arsenic in them, but metallic tin is pure tin, so definite amounts of tin can be added each time.

Now you have your ingredients and your recipe for bronze, it is time to make some.

FROM STONE TO BRONZE

Arrows were one of the most powerful weapons in the ancient arsenal. Light, easy to use, and able to strike at a distance, they gave hunters and soldiers a huge advantage over their prey, be it human or animal. In the Stone Age the arrowheads of choice were often made from knapped stone.

By striking certain stones like flint at an angle it is possible to create an incredibly fine edge that can cut through flesh with ease. Many different shapes could be made that altered the behaviour of the arrow in flight and when it struck its target. Throughout the Stone Age, and well into the Bronze Age, these arrowheads were the favoured way to tip arrows.

Bronze, however, slowly pushed stone aside. Stone arrowheads can be fearsomely sharp yet the brittle material they are made from would shatter against bronze armour. To defeat an armoured enemy therefore, bronze arrowheads were needed. While it may have taken a skilled knapper a long time to craft a single arrowhead, many bronze ones could be turned out of moulds quickly. In the brutal world of bronze, stone simply could not cut it any longer.

Egyptian bronze arrowheads like this one could punch through armour and into the body of an enemy without bending or breaking

© Rama

Molten bronze flows to fill the shape of whatever mould it is poured into which allows for easy manufacture of complex items

Introducing new elements to a metal may disrupt its atomic structure and produce a material with radically different properties

Making Bronze

Mining and smelting copper had given Bronze Age peoples a good handle on the technology required for producing bronze. Tin melts at a temperature of 232°C, far below the melting point of copper, which is 1085°C. Once copper has been melted in a furnace, adding the necessary amount of tin to the liquid copper will automatically produce molten bronze. This could then be poured out into ingots or used immediately to cast other objects. Getting tin may have been difficult in the Bronze Age but making use of it was easy.

One of bronze's key abilities is that it can be easily melted down and recast for use again. If a sword breaks or armour is broken then it can be returned to the furnace to be recycled. Because bronze becomes a liquid at 950°C, below the melting point of copper, no special heating technology was required to reuse bronze. A smith with a pair of bellows would soon have bronze reduced to its liquid form in a furnace.

To cast bronze several techniques were used. Perhaps the most simple was that of sand casting. By creating a wooden frame and filling it with sand you can create a mould by impressing whatever object you wish to make into it. You could make a model of an axe out of clay, for example, and push it into the sand. Then a second frame could be placed on top and also filled with sand. By compacting the sand as closely as possible the frame can then be opened, creating a hollow the exact same shape as the desired object. By carefully removing sand to create a channel for the molten bronze to enter, you have now got everything you need to cast a metal object. Given the ephemeral nature of the tools used, however, there is little evidence of sand casting in the archaeological record.

Stone moulds were commonly used in many locations and have survived. From Bronze Age Italy stone moulds have been found for the manufacture of axes, daggers, sickles, and ornaments. The process of making a stone mould involved breaking fine-grained stones in two with a bronze

The Houmuwu Ding is the largest bronze item surviving from the Bronze Age and was produced from a clay mould

©Wiki/ Mlogic

chisel and carefully excavating the shape of the final object. This was a skilled technique and would have required a great deal of practice and patience to create moulds, but once they were made a stone mould could be used over and over again to cast hundreds or thousands of identical bronze products.

Unlike a stone mould, a clay one can be used only once to create a bronze object. Using a clay mould could be a simple act, such as covering a blade to create a mould for a sword, or it could be highly complex. In the late Bronze Age the 'lost wax' technique was used, where a model would be made of wax and surrounded in clay. As the clay was fired in a kiln to harden it the wax would burn away to leave a hollow mould.

In China bronze containers called ding were made from clay moulds that involved many complicated steps. First a model of the container, including its decoration, would be made from clay. Once this had hardened then wet clay would be pressed onto it to create a negative impression. These negatives would be allowed to dry until they could be cut off. The original clay model was then shaved down by the desired thickness of the final casting. By reassembling the negative clay panels around the now smaller original, a highly decorated chest could be cast. The Houmuwu Ding, dating from around 1000 BCE, weighs 830 kilograms and is the largest bronze item surviving from antiquity.

Bronze's unique advantages over the copper it replaced were its durability, strength, and ability to hold a sharp edge. To make a blade as sharp as possible from bronze, however, requires it to be work-hardened. By beating the edges of a blade, imperfections from the casting process can be removed, but it also makes changes at the atomic level. Beating creates smaller grains within the metal and, much like alloying itself, this creates less chance of planes of atoms within the metal sliding past each other.

The final step in creating bronze objects is polishing them. A polished bronze, depending on the alloy used to make it, can appear as either gold or silver - though of course much stronger than either of those two glittering metals. The shine of bronze would have made it one of the most high-status materials available at the time.

Using Bronze

It is hard to be certain what the first use of bronze was. While dating metals can be problematic there is also the fact that bronze was a highly valuable substance in the ancient world. Only the most exalted people were likely to be buried with metal objects. Bronze, which can be reworked almost infinitely, was probably recycled rather than wasted. Among the earliest bronze artefacts discovered are weapons like spearheads, arrowheads, and daggers. There are also a great many examples of bronze being used for art, jewellery and decoration.

Polished bronze glows by daylight and shimmers in the light of the wavering flames of lamps. Bracelets and other items of jewellery made from bronze have been discovered across the world from Bronze Age sites. A bronze statue known as 'The Dancing Girl' from the Harappan civilisation of the Indus Valley around 2600-1900 BCE shows a female figure decked out with 29 bangles, 25 of them on her left arm. The beauty of bronze should not be forgotten among its more mundane or murderous uses.

It was in weaponry that bronze had its most profound effect. Bronze's ability to hold an edge opened new frontiers in warfare. Stone weapons, maces, and knapped arrowheads had dominated weaponry before the invention of bronze. Pure copper is too soft to be of much use in a battle where it would soon be dulled or bent. Bronze however could withstand multiple strikes and still be a deadly threat.

Daggers of copper and bronze are found in Egypt from around 3100 BCE but appear to have been mostly ceremonial and were often highly adorned. While short daggers would have been practical weapons for very close fighting in a large, massed battle, they must have been the choice of last resort. Spears, arrows, and axes were the predominant force on the battlefield. Daggers would only come into their own when they were slowly lengthened into the blades that we call swords.

There is no clear definition of what a sword is and what a dagger is. At Arslantepe in Anatolia a tomb was found that contained either long arsenic bronze daggers or short swords from around 3000 BCE. All were cast in a mould and several bear intricate silver decoration. It was only around 1800 BCE however that the first swords truly developed elsewhere in the Mediterranean. Almost immediately thereafter swords are found across the Middle East and high in northern Europe. In a find from 1600 BCE in Nebra, Germany, two bronze swords were recovered, showing how quickly knowledge of sword creation had spread.

The earliest bronze swords had no tang, a piece of metal protruding from the bottom of the blade, and so had to have their handles riveted on. This introduced a weak point which must have left many warriors in battle holding just a handle after the blade had been sheared off on impact. Many early bronze swords have been found that show rivets having been torn out. Later swords kept the rivets even after a tang had been added to the blade which removed the chance of the sword flying off.

Of course bronze was not only used in warfare and for decoration. It would always have been a costly material, but it did find a place in the farm and the home. Bronze axes could cut down trees as easily as they could men. Ploughs with stone blades did a fine job but if they struck another stone in the field then they often shattered. With a bronze plough the damage, if any, could be repaired quickly. When harvest came a bronze sickle could make light work of reaping grain.

Bronze could make a huge difference in even small villages. In Ban Chiang in Thailand many artefacts have been discovered from ~1900 BCE that reveal a whole array of uses that did not rely on warfare. Ban Chiang was not part of a large bellicose society and yet they still imported and worked with bronze to produce tools like adzes for shaping wood, kitchen utensils, and decorative bangles and bells.

The Bronze Age may have changed the world through trade and on the battlefield, but for most of the people at the time it was the subtler, quality-of-life improvements in their lives that were brought about by bronze that would have most impressed them.

The Dancing Girl sculpture from the Indus Valley shows how bronze was used to create objects of beauty, as well as destruction

©Wiki/Gary Todd

SWORDS OF THE BRONZE AGE

As soon as swords developed they began evolving into differing forms as people sought ever more efficient ways of cutting down their foes. The first swords were used much like daggers, with sharp points most suited for stabbing into enemies. With a weak spot at which the blade connected to the handle, slashing with them risked a sword breaking apart in the hand.

Swords with wider blades that were better able to survive battle, known as leaf-shaped blades, developed. With sharpened edges on both sides of the swords a warrior could hack at soldiers in any direction. A composite sword type known as the carp-tongue featured a triangular blade that could be used both for stabbing and slashing.

In Mesopotamia the curved sickle sword called the khopesh was favoured. Only the outer edge of its hooked shape was sharpened to a cutting edge. A thicker inner edge gave the blade strength while the tip could still be used to drive into an enemy. With the curve a soldier could hook and pull away at a shield to strike their opponent. The khopesh became the sword of choice in a number of Middle Eastern cultures before being abandoned at the end of the Bronze Age.

Bronze Age swords were used and often broken in battle but were also highly ornamented to show the value placed on their lethal power

THE BUILDING BLOCKS OF SCIENCE

SCIENCE DID NOT SPRING FULLY FORMED INTO HISTORY. RATHER, ITS FOUNDATIONS WERE LAID THROUGH THE MILLENNIA AS PEOPLE IN THE PAST, NO LESS INTELLIGENT THAN US TODAY, SOUGHT TO UNDERSTAND AND CONTROL THE WORLD AROUND THEM

WRITTEN BY **EDOARDO ALBERT**

For most scientists the scientific worldview is the very air they breathe. They are no more aware of it than we are normally aware of the air that surrounds us. However, the whole scientific enterprise depends upon deep foundations. Some of these foundations are very deep indeed, going back to before modern humans, homo sapiens, even existed. Others are more recent in historical terms but are nevertheless still thousands of years old. And finally, there are the specific ideas that made the rise of science possible within a particular culture and at a particular time.

Some of these advances are basically technological, such as the discovery of iron smelting. There were other vital technological advances, ranging from the first stone tools to the improvements in lenses that made telescopes possible, that became huge disruptive advances in the way that people interact with the world.

But even more important than the technological advances were changes in the way people thought about and understood the world. Prehistoric cultures lived in a thought universe of indwelling spirits and presences which could be appealed to and affected by various rites and rituals. This was a world where capricious deities ruled the elements and fate was inexorable but also unpredictable. This worldview continued into historic times among the Greeks and the Romans. Such beliefs make science impossible to even conceive.

But during these times there were changes afoot, which were philosophical and cultural, that began to make the process of science imaginable. So we will also look at how various thinkers changed the way we view the world enough for science to become imaginable.

This is a story of advances and strange revolutions, of anonymous inventors and monks on the edge of the world. It is a story of humanity looking out into a strange and oftentimes hostile universe and trying to understand just what on Earth is going on. That you can read this now is a testament to the success of all these people who went before us. They made it all possible.

"EVEN MORE IMPORTANT THAN TECHNOLOGICAL ADVANCES WERE CHANGES IN THE WAY PEOPLE THOUGHT ABOUT AND UNDERSTOOD THE WORLD"

THE WEAPON OF THE GODS

OTHER ANIMALS USE TOOLS. IT'S BECOMING INCREASINGLY CLEAR THAT OTHER ANIMALS HAVE SOPHISTICATED LANGUAGES. BUT FIRE, ITS TAMING AND ITS USE, IS UNIQUE TO HUMAN BEINGS. IT MADE US GODS

We still live in the Iron Age: Space X's Starship is made of stainless steel and this is the vessel that will take us to Mars

THE HARD STUFF

Stone has an obvious advantage. There's a lot of it about. There's also a lot of wood. So for thousands upon thousands of years, human beings were content with tools made from stone and wood. But around 3300 BCE, people in Mesopotamia discovered that if you melted copper, which has a relatively low melting point of 1084 degrees Centigrade, and mixed it with tin, then when the mixture cooled down it produced this marvellously hard and brightly coloured substance, bronze. Bronze could be polished. It could be worked. It could be sharpened. And everyone who was rich and powerful wanted it. But unfortunately, while copper is common, tin is not. There were tin mines in Cornwall, in Spain and poorer sources in Brittany and Italy, but not much else. This limited the use of bronze. However, about 2000 years after the discovery of bronze, metal workers in the Middle East learned how to make hotter smelts that melted another rock which was often found with copper deposits. This was iron in various forms of oxide, and it is one of the commonest elements on earth. Being common, iron tools and weapons were not confined to an elite, but could be used by ordinary farmers and to equip armies. The Iron Age had begun. And it is still going on.

Legend says that long ago in the morning of the world, the great god, Zeus, gave to the titans, Prometheus and Epimetheus, the task of making creatures for the world. Epimetheus made the animals, giving them great gifts. Some could fly and others swim without drowning. To some he gave great strength and size, others he made small that they might hide.

Prometheus made a creature also, and this he made, in form and manner, like to the gods. But when he sought to give it gifts he found that Epimetheus had given all the gifts away. These creatures he had made had neither great strength nor armour, they did not have sharp teeth and claws, nor could they fly or run as fast as the other creatures. They were naked, helpless and defenceless, left shivering in the cold of the world.

Seeing the creatures he had made so vulnerable, Prometheus took pity on them. He decided to give them something by which they could defend themselves against the other creatures his brother had made. So one night, Prometheus climbed Mount Olympus, where the gods lived in splendour and ease. There, he crept into the forge of Hephaestus, the smith god, and from the fire there he took a single spark, hiding it in the hollowed-out stem of a fennel plant he had brought with him.

Taking the spark of fire, Prometheus crept from Olympus and slipped back down the mountain. He went to the creatures he had made and gave to them the gift he had brought: fire. But having given them fire, Prometheus taught his creatures other skills: the forging of metal and how, by making marks on clay or reed, they could catch words and save them. Now those helpless creatures, made in the image of the gods but with none of their powers, were not so helpless any more.

However, when Zeus saw what these creatures were doing, how they waged war against each other and swarmed over the world, paying little attention to the gods, he became enraged. He trapped Prometheus and bound him to a rock. There, each day, an eagle came and pecked the liver from the titan only for it to regrow each night so that the punishment might begin afresh when the sun rose on the next day. This is the myth of Prometheus, first recorded in Hesiod's *Theogony*, written in the 8th century BC. There are related stories of the theft of fire from the gods in other mythologies around the world. These are among the oldest stories we have and they show how, from the first, our ancestors knew that the secret of fire was something that set us apart from other creatures. As King Louie, the king of the monkeys, says to Mowgli in *The Jungle Book*, "Give me the power of man's red flower/So I can be like you."

The irony is that the use of the red flower predates modern humans.

It is, of course, difficult to find incontrovertible evidence of the intentional use of fire. By its nature, fire doesn't leave many traces and fires also occur naturally following lightning strikes, eruptions and other events. However, archaeologists have discovered the remains of a hearth and large amounts of burnt bone from a cave in Israel. The cave was inhabited for 200,000 years, between 400,000 and 200,000 years before the present - an unimaginably long span of time.

"FIRE MAY HAVE TRIGGERED EVEN MORE PROFOUND CHANGES THAN EARLIER IMAGINED"

The people who lived in Qesem cave were homo sapiens. But in a cave in South Africa, archaeologists have now found clear evidence that our evolutionary ancestor, homo erectus, also used fire. The location of these finds is Wonderwerk cave in Northern Cape Province. The cave is deep, pushing 140 metres (460 feet) into the hill that rises above it. Deposits have been found in the cave dating back as far as two million years. Within one deposit, containing stone flakes and hand axes, archaeologists found charred bones and the ash residue of plants.

The extraordinary thing is that this layer has been dated to our ancestor, homo erectus. Homo erectus lived between two million and 200,000 years ago. While recognisably human-like, homo erectus, dressed up in modern-day clothes, would not easily pass for homo sapiens. There is also still some debate among scientists as to whether his physiology allowed for complex speech sounds. But whether homo erectus could speak or not, he could make fire.

The mastery of fire may also have triggered even more profound changes than earlier imagined. For there are interesting evolutionary changes that occured following the discovery of the use of fire. Teeth and stomach became smaller while the skull got bigger. We know from the remains at Wonderwerk that homo erectus started cooking food, roasting meat and cooking plants. Cooked food is softer and easier to chew, allowing for the evolution of smaller teeth. Softer food also means that the jaw muscles do not have to be so strong, allowing for an expansion in the size of the skull. And preparing food by cooking allowed for a wider range of food to be eaten, while smoking allowed food to be preserved and eaten when times were lean. So the use of fire in cooking may have played a key role in the eventual evolution of our own species.

Fire allowed our naked forebears to withstand temperatures that might otherwise have killed them, as well as keeping predators from night-time camps. And the natural impulse to gather in a group around the fire, staring into its depths, might well have played a part in the evolution of language and social bonds so vital to human development.

At some point, our ancestors also learned to use fire to change the landscape by controlled burning. As bush fires are a natural phenomenon, it's impossible to say when directed human use of controlled burning began but it's likely to be ancient too, as our ancestors could easily observe and follow the effects of natural bush fires.

Given all this, it's no surprise that the discovery of fire figures in so many mythologies. In terms of its effect on human development, fire ranks as probably the single most important discovery in human, and pre-human, history.

Thank you, Prometheus. We owe you.

GREEK FIRE

THE ANCIENT GREEKS WERE THE FIRST TO LOOK AT THE WORLD AROUND THEM AND ASK HOW IT WAS ALL HAPPENING

We live now in a stable world. Things don't change much. Most of us can count on living to old age. For about the last century, parents can reasonably expect to see all their children grow to adulthood.

It wasn't always like this. To the people living in the small Greek city states that had pulled themselves out of the civilisational collapse which had ended most of the cultures in the area during the 12th century BCE, everything seemed in flux. Babies were born and died, many before they could reach adulthood. Cities and civilisations rose and fell. Famine followed a year of rich harvest. And this human impermanence was set against a backdrop of inexplicable nature.

Most other cultures explained this impermanence in theological terms. But the Greeks, while they had their epic poems, the *Iliad* and the *Odyssey*, had no theology to compare with that of Egypt or Mesopotamia. By contrast, the Greek gods, who mingled all too freely with human beings according to the tales poets told of them, were little better than hormone-driven teenagers, randy as hell and with less self-control than toddlers, set within immortal bodies. They might be placated or appeased, as you might placate or appease a mercurial king, but there was nothing there to satisfy the deep human need to understand the world.

Without a theological explanation for the world around them, thinking Greeks had to look elsewhere. The first to do so was Thales of Miletus, who lived in the 6th century BCE. Thales was the first natural philosopher, for he looked for explanations of the world around him in natural processes rather than ascribing everything to the direct action of the gods.

While Thales saw a world of change, he also discerned that there was an underlying constancy under this change. Thales argued that this was because there was a single substance that, by changes of state, brought all other substances into being. According to Thales this universal substance was water, which we can easily see in different states as a solid, liquid, and gas.

Thales further argued that everything had an inner divinity that ordered it towards their proper end. Thus Thales proposed a natural world that was ordered towards particular ends and that the world was not a mechanism but an organism. This emphasis on the ends towards which things were directed, teleology, was fundamental in ancient Greek thought.

Thales inadvertently helped develop another key element of the scientific method: the critique of existing theories. His own pupil, Anaximander, pointed out that the fundamental quality of water is wetness. Water can't be water and not be wet. But as there are substances that are fundamentally dry, water cannot be their underlying substance, for then water would be contradicting its own essence.

> "GREEK NATURAL PHILOSOPHY REACHED ITS PEAK WITH ARISTOTLE AND ARCHIMEDES"

This set off a long argument among natural philosophers which eventually culminated in Greek thinkers accepting that there were four basic elements that combined to make everything else: earth, air, fire and water.

While this accounted for the substance of things, it said nothing about what form they took. This problem was tackled by Pythagoras (of theorem fame), who imported mathematics into natural philosophy based on his belief that numbers were the fundamental structure of the world.

Greek natural philosophy reached its peak with the works of Aristotle and Archimedes. Aristotle analysed the world around him according to four causes: the material, the formal, the efficient and the final cause: what something is made of, what is its form, how does it acquire that form and, most importantly, to what end is it directed, what is its purpose.

For Aristotle, the natural spontaneous operation of things was what counted, and this is what he observed (he was a superb observer of biological life). Carrying out an experiment meant interfering with the natural operation of a creature or a system, and thus spoiled its natural operation. So

A satellite view of Greece, showing the Peloponnese peninsula and the narrow isthmus of Corinth

experimentation was ruled out in his world view. However, mathematics fitted this view of things well, and Aristotle made important advances in maths.

While Aristotle was content to observe, Archimedes was as close as classical antiquity came to a true experimenter, as well as being a brilliant mathematician. He analysed how cones can be cut into sections as well as ways to calculate the area of a circle in such depth that later mathematicians saw his work as the heart of their own research into calculus.

Archimedes was a citizen of the Greek city state of Syracuse on present-day Sicily. Much of his work was directed towards the improvement and protection of Syracuse, at that time coming into the sights of the expanding power of the Roman Republic. The king commissioned Archimedes to design a ship for him and he responded with the Syracusia, the largest ship of the ancient world, complete with gym and temple to Athena. To pump out water from the bilges, Archimedes designed a revolving screw inside a tube that is still in use today to raise water from one level to another.

During the long siege of Syracuse (spring 213 BCE - autumn 212 BCE) he invented devices to protect his home against Roman forces. Among these was a giant grappling hook that swung out from battlements, catching Roman ships, lifting them from the water and tipping them over. According to later reports, Archimedes also set up a system of mirrors to focus reflected light on the attacking ships, making them burst into flames, although modern attempts to recreate this with the technologies available to Archimedes have largely failed. Sadly, Archimedes' attempts to stop the Romans failed, and he was killed when the city fell.

The great achievement of Greek thought was to make the world an object of rational inquiry. Before them, the world was an arena of competing and unpredictable supernatural creatures. You might buy off threats or appeal for something from the gods, but you could not predict what natural phenomena were going to do. The Greeks imported order and predictability into the world. But it was still a world that had to conform to the perfections of the world above forms, the eternal forms of mathematics and Platonic theory. To confront the world in its actuality required a fresh inspiration. That inspiration would come from an unexpected source.

DIOLKOS – THE ANCIENT GREEK RAILWAY

If you look at the map of Greece, you'll see that it has two, three-fingered hands that reach into the Mediterranean Sea. The eastern one is the Chalkidiki peninsula, the western and larger peninsula is the Peloponnese. The wrist joining the Peloponnese to the mainland is the narrow isthmus of Corinth. This isthmus is only four miles wide and the Greeks quickly realised that, if boats could get across the isthmus, they would save themselves a long and often dangerously stormy voyage around the headlands of the Peloponnese. So they built a railway. The Diolkos was a tracked road right across the isthmus of Corinth that operated for over 600 years, until the middle of the first century CE. Ships were apparently hauled out of the sea and onto wooden rollers before being pushed onto a wagon with wheels that ran in grooved tracks. It took huge teams of men to haul a ship across the isthmus so in normal times the Diolkos may have been more often used to move a ship's cargo across the isthmus to another ship waiting in port at the other side. But ancient sources record warships, triremes, being hauled over the Diolkos in times of trouble, so ships did definitely make the overland journey too. It was the world's first public railway.

THE NOT-SO DARK AGES

MANY OF THE FOUNDATIONS FOR THE BIRTH OF MODERN SCIENCE WERE LAID IN THE CENTURIES FOLLOWING THE COLLAPSE OF THE ROMAN EMPIRE, DURING THE PERIOD OF INTELLECTUAL FERMENT AND GROWTH KNOWN AS THE MIDDLE AGES

In 725, a monk living in a wind-swept monastery on the edge of the world sat down to write a book about time. The monk's name was Bede and he had been a boy when the monastery was founded. He had spent all his life living there, in Jarrow, with trips to its brother monastery seven miles down the coast at Wearmouth. The twin monastery of St Peter's, Wearmouth, and St Paul's, Jarrow, was blessed with what was probably the best library in western Europe. Its contents had been painstakingly assembled during the course of five trips to Rome by the monastery's founder, Benedict Biscop.

Bede had benefited immeasurably from this library, learning fluent and elegant Latin, reasonable Greek and even some Hebrew. Bede's own tongue was Old English and, two generations earlier, his grandparents had been illiterate pagans. But the Anglo-Saxons had embraced the new world of learning that opened up to them with their conversion to Christianity and exposure to books written in Latin, which was the international language of the Church.

Nowhere was this learning more enthusiastically espoused than at Bede's own monastery. The monks embarked on producing a definitive, scholarly edition of the Bible, with much of the critical work done by Bede himself, and, extraordinarily, one of the three editions they produced still survives: the *Codex Amiatinus*, now held in Florence.

To produce this critical edition of the Bible, Bede and the other monks had to compare and contrast different texts, work out where errors may have been introduced by faulty translations and scribal errors, and correct those mistakes. The undertaking was an extraordinary feat of scholarship.

And, critically, it was indicative of a crucial change in how medieval scholars looked upon texts from the past.

Ten years after his work on the *Codex Amiatinus*, Bede wrote another book, *De*

"SCIENCE ONLY BECOMES POSSIBLE IF IT IS IMAGINABLE. THE WORLD HAD TO BE DIS-ENCHANTED"

Temporum Ratione ('On the Reckoning of Time'). Although little read now, it was hugely influential in the centuries after it was written, being copied and recopied across the monasteries of Europe.

The book, as the title suggests, was about time. In it, Bede looked at the various natural expressions of the flow of time, such as the lunar and solar cycles, but he also considered the tides. Living near the mouth of the River Tyne, a couple of miles inland from the North Sea, Bede was acutely aware of tides. At North Shields, the Tyne has a normal tidal range of five metres (15 feet).

Looking out of his door, Bede saw something very different to what his classical sources said about tides. In those much revered books, he found authors describing tides that shifted barely at all. What's more, some of these ancient authorities also averred that the tide rose and fell everywhere at the same times.

Bede was part of a monastic network. There were other monasteries up the coast at Lindisfarne and down the coast at Whitby, separated by a hundred miles. Bede contacted his brethren at these monasteries and asked them to record the times of high and low tides at their respective monasteries. The difference in tide times between Lindisfarne and Whitby is between an hour and 90 minutes. This was a big enough difference to be measurable.

So Bede noted what his classical source wrote in his book and then compared what antique authority said about the tides as compared to the evidence he had accrued - and found the classical authority wrong.

This was a crucial step in the development of science. Bede had put elegantly expressed classical theory, with all the weight of prestige that it carried from its association with the might of Rome, to the test of empirical evidence - and given the greater weight to the empirical evidence.

This was a huge change from classical thought. For all its depth, classical thought was essentially bound to an idealised version of reality. The circle was the perfect shape so therefore the planets must move in that perfect shape. If there was a clash between philosophical truth and empirical evidence, then it was the evidence that was wrong.

Christian thinkers changed that. The world, they argued, was created by a rational God who had made a rational world. It was a world that obeyed the laws that God had laid upon it. What was more, by understanding the laws of the natural world, it was possible to understand the nature of its creator better - so there was a religious incentive to investigate the natural world and understand it.

There is no self-evident reason why the world has to make sense, still less that it should be understandable to human reason. While there are recurring features, such as the days and the seasons, there are many more that seem random and inexplicable, from the roaring of the thunder god to the rising of the sea god. But Christianity had banished these nature gods: the dryads and nymphs of the Mediterranean, the elves and wights of the northern forests, were all officially dismissed. One God alone made everything and, unlike in the Islamic world where God directed everything by direct acts of his will, in Christian thought he moved the world by his laws.

Science only becomes possible if it is imaginable. The world had to be dis-enchanted before it could become explicable. And theory had to bow to the test of evidence. Both of these are absolutely fundamental prerequisites for the possibility of true science, and this shift in worldview was slowly accomplished in the centuries that later, more scornful times, labelled the Dark Ages. But as scholars are slowly establishing, these centuries were not so dark after all, and not least among their accomplishments was the laying of the intellectual foundations for the rise of science as we know it today.

FREEDOM

The centuries between the fall of Rome and the Renaissance were a time of extraordinary, if largely unreported, technological and cultural progress, from the three-field rotation system, which meant that medieval peasants were much better fed than Roman peasants, to tidal- and windmills. A key contributory factor to this progress was that there were no more slaves. The Church had largely succeeded in ending slavery in Europe by the Middle Ages. This was a key difference to the Roman Empire, and indeed most other societies around the world. Any problem can be solved if you throw enough people at it. Widespread slavery gave rulers and the rich enough slaves that they did not need to look for labour saving solutions to any problem. But with the ending of slavery, endless free labour was no longer available. It became necessary to find other ways to do things. Among the fruits of this search were tidal- and windmills, which used natural forces to grind wheat rather than having slaves do the labour, and better harnesses which meant that horses could pull ploughs rather than oxen or men having to do so. On the cultural level, one of the biggest obstacles to the technological innovations needed for science was a slave-based economy. By the Middle Ages, that had been abolished in Europe. Another of the building blocks was laid in place.

The Roman economy was slave based and, therefore, extremely inefficient, for slaves have no incentive to work other than to avoid a beating

"BY 300 BCE, THEY HAD DEVELOPED MATHEMATICAL MODELS TO DESCRIBE THE MOTION OF THE PLANETS"

The entrance to Newgrange. At the winter solstice the rising Sun shines through the roofbox, the hole above the door, illuminating the main passage

Ziggurats were great platforms from which Babylonian priest-astronomers could carry out their observations for generation after generation

THE FIRST ASTRONOMERS

HOW THE ANCIENT BABYLONIANS DEVELOPED THE FIRST TRUE SCIENCE

WRITTEN BY **EDOARDO ALBERT**

The movement of the Sun and the Moon are fundamental to human life - indeed, all life on Earth. That our early ancestors were aware of these movements and that they tracked them is clear from ancient sites such as Newgrange in Ireland, where the rising Sun at the winter solstice shines down the main passage, and objects such as the Nebra sky disc, which depicts the Moon in full and crescent phases as well as the Pleiades.

However, it was in Mesopotamia, the land between the rivers (modern-day Iraq) that astronomy became the first field of human knowledge to approach the level of a science: by 3000 BCE, astronomers from the city-state of Babylon were able to predict planetary behaviour, such as when Jupiter's apparent motion in the sky would go retrograde (this is when planets appear to move backwards in the sky relative to the fixed stars).

For the ancient Babylonians, the movement of the Sun, Moon and planets were intimately linked to the fortunes of the king and his kingdom: as above, so below. So the priestly caste, using the ziggurat temples as their observational platforms, began keeping extensive records of what they saw happening in the sky. The Babylonians had also invented a written alphabet, cuneiform, which they recorded on clay tablets. Since these clay tablets are virtually indestructible so long as they are kept dry, Babylonian astronomers soon amassed a huge database of their continuing celestial observations.

Amassing observations over a period of years soon showed the Babylonians that some celestial patterns repeated. While they did not have the ability to make precise observations, perseverance more than made up for this. Within a few decades, Babylonian astronomers discovered that it was possible to predict the movements of the Moon and planets. To learn this, records spanning decades were vital. Neither planets nor the Moon repeat their cycles annually; the cycle is longer than that. For instance, it takes eight years to see Venus go retrograde at the same point in the sky. To observe Mars repeating its celestial behaviour, they had to watch for 47 years, while Saturn takes 59 years.

By having such a record, Babylonian astronomers could predict the motion of Venus by looking to see what it was doing eight years earlier, while for Mars and Saturn they would have needed to look at each planets' celestial diaries 47 and 59 years previously.

But Babylonian astronomy did not rest there. By 300 BCE, they had developed mathematical models to describe the motion of the planets that allowed them to predict their motion into the future.

As such, ancient Babylonian astronomy can be called the first true science, for it made predictions about the future behaviour of celestial objects that could be verified by their actual motion. And as a by-product, they also decided that there should be 360 degrees in a circle, roughly mapping the days of the year onto the round of the sky.

THE CRUCIBLE OF ALCHEMY

ALCHEMY BEGAN IN THE UNIQUE SURROUNDINGS OF 3RD- CENTURY ALEXANDRIA. WHAT WAS IT ABOUT THE EGYPTIAN PORT CITY THAT MADE IT THE BIRTHPLACE OF THIS MYSTERIOUS ART?

WRITTEN BY **APRIL MADDEN**

Today there's a clear delineation between the concepts of religion, magic, and science, but this is a relatively recent invention. In the 3rd century Roman province of Aegyptus, however, such a distinction wasn't made. The world was understood on the basis of empirical evidence filtered through the philosophies and mythologies of a wide range of religions and cultures, and from this unique, curious mixture rose a new way of seeing: alchemy. Some historians may try to argue in favour of a purely exoteric beginning for alchemy - that its roots are entirely material, artisanal, and proto-scientific - while others prefer focusing on its esoteric aspects, its myth and mysticism and even magic. The division between the two is a modern one, however, coined by British writer Mary Anne Atwood in the 19th century. 1700 years ago, the multicultural artisans of Alexandria, the capital of Aegyptus, didn't split ideas into hypothetical versus substantial applications or divide them into sections labelled real, fanciful, and downright made-up. Instead, they encompassed them all into one interconnected, holistic, practical yet arcane worldview that was full of possibilities. It's this worldview that radically influenced how alchemy began, and it could have only arisen the way it did in the unique society of Alexandria.

Situated in the north of Egypt, on the Mediterranean coast and the now defunct Canopic branch of the Nile, then a sizeable channel capable of supporting the deepwater ships that ferried cargoes of metal in and out of the small port town of Rhacotis, the capital of Graeco-Roman Egypt was constructed under the auspices of its 4th century BCE namesake, Alexander III of Macedon (Alexander the Great), who shortly after conquering Egypt dreamt about his favourite poet describing this little western corner of the Nile Delta to him. The young Greek king travelled there clutching his treasured copy of the works of Homer and wandered around until, standing on the rocky shore of the nearby island of Pharos, he was inspired to build a city.

The newly named Alexandria was designed from scratch by the architect

Alexandria was symbolised by its iconic lighthouse, the Pharos, one of the Seven Wonders of the Ancient World. As well as keeping shipping safe, it was reputed to represent the city's status as a beacon of knowledge

A diorama depicting a typical Alexandrian khēmia workshop, based on archaeological remains found in the city and manuscripts describing laboratories

"ALCHEMY IS THOUGHT TO HAVE BEGUN AROUND A CENTURY BEFORE THE ROMAN BAN"

Dinocrates of Rhodes, an unparalleled opportunity for a town planner who'd previously only been able to insert the odd new build or standalone district into much older, more organic, sprawling urban centres. Legend has it that, lacking chalk, Dinocrates laid out his life-size gridiron plan for the city using barley flour. The problem was, the local seabirds kept devouring it while Dinocrates and his workers were trying to take the necessary measurements from the drawings. Many of them were superstitious, claiming that the birds' continual destruction of the designs was an omen that Alexandria shouldn't be built. Alexander's personal soothsayer, Aristander, however, took the birds' hunger as an augury that the new city would nourish the entire world. This interpretive tension between the preternatural and the pragmatic was at the heart of the next several centuries of Alexandrian life, centuries that eventually saw the unique Graeco-Egyptian city come under imperial Roman rule.

The Roman conquest of Egypt was the result of the death of Pharaoh Cleopatra VII Philopater in 30 BCE after a short, bloody war prosecuted by the Roman statesman Octavian (later Emperor Augustus Caesar), due in part to the political fallout from the Egyptian queen's tactical but ill-fated love affairs with first Octavian's adopted father and benefactor, Julius Caesar, and then his later political rival Mark Antony. Cleopatra herself wasn't strictly

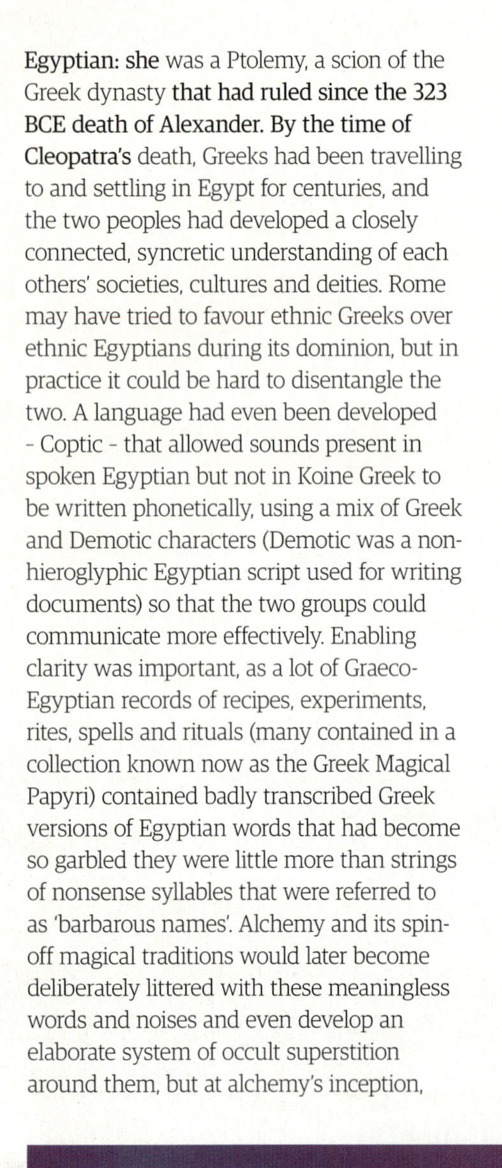

Egyptian: she was a Ptolemy, a scion of the Greek dynasty that had ruled since the 323 BCE death of Alexander. By the time of Cleopatra's death, Greeks had been travelling to and settling in Egypt for centuries, and the two peoples had developed a closely connected, syncretic understanding of each others' societies, cultures and deities. Rome may have tried to favour ethnic Greeks over ethnic Egyptians during its dominion, but in practice it could be hard to disentangle the two. A language had even been developed - Coptic - that allowed sounds present in spoken Egyptian but not in Koine Greek to be written phonetically, using a mix of Greek and Demotic characters (Demotic was a non-hieroglyphic Egyptian script used for writing documents) so that the two groups could communicate more effectively. Enabling clarity was important, as a lot of Graeco-Egyptian records of recipes, experiments, rites, spells and rituals (many contained in a collection known now as the Greek Magical Papyri) contained badly transcribed Greek versions of Egyptian words that had become so garbled they were little more than strings of nonsense syllables that were referred to as 'barbarous names'. Alchemy and its spin-off magical traditions would later become deliberately littered with these meaningless words and noises and even develop an elaborate system of occult superstition around them, but at alchemy's inception, the people of Graeco-Roman Egypt were focused on sharing and revealing rather than obfuscating their knowledge.

And those people weren't just Greek, Roman and Egyptian. Aegyptus was a world-famous cultural melting pot, a crossroads situated between Mediterranean Europe, Africa, and the Near and Middle East. Before Alexander's invasion, Egypt had, by and large, been under the purview of the Persian Achaemenid Empire since 525 BCE. The early Persians didn't impose cultural homogeneity on their conquests: Egypt had been permitted to keep its traditional religion and society intact, but Persia's scholarly Zoroastrian clergy would have certainly investigated its wealth of knowledge as well as sharing their own highly influential ideas (and those they'd picked up from forays as far afield as India and China). There were still plenty of ethnic Iranians in Egypt's rich cultural mix even after Alexander's conquest, followers of Zoroaster or of the upstart Sasanian preacher Mani and his syncretic 'Religion of Light', which drew influences from Judaism, Buddhism, Hinduism, Christianity and Zoroastrianism together into a faith

This idealised, Baroque 18th century history painting of Alexander the Great founding Alexandria by Placido Costanzi depicts the city's foundation as divinely inspired. Architect Dinocrates is shown demonstrating the city plans to the king

GODS OF ALEXANDRIA

Alexander the Great himself was such an inspiration to the ancient world that even before his death he was viewed as a demigod when he conquered Egypt - he was crowned as Pharaoh in Memphis and named 'Son of the gods', which affected the religious young king very deeply. After his death his remains and even his legend were venerated.

The dynasty that ruled Egypt after him, founded by his companion and possible half-brother Ptolemy I Soter, recognised that blending the gods of Egypt with those of Greece would go a long way towards unifying a population that had only come together thanks to its adoration of Alexander. The Ptolemies began a syncretic process of combining the gods of both cultures: Zeus-Ammon was already a duality both societies were comfortable with; to that they added Dionysus-Osiris, Isis-Aphrodite, Hermanubis, Hermes Trismegistus, and, most importantly for Alexandria, Serapis, to whom was dedicated the Serapeum of Alexandria, a temple complex that was referred to as the "daughter of the Library of Alexandria", suggesting that it too was a centre of learning. Serapis was a synthesis of two Egyptian gods, Osiris and Apis, but represented with entirely Greek iconography, as a strong, middle-aged man in the prime of life, akin to Zeus, with chthonic and water-god elements.

Alexandria's patron god Serapis was a combination of Egyptian fertility god Osiris and sacrifice and rebirth god Apis, with Greek iconography depicting him as a strong, patrician pater familias for the entire city

known as Manichaeism. From the south, travellers and settlers from the Black Land's ancient African allies in what are now Eritrea, Djibouti, Ethiopia, Somalia and Sudan either followed similar pantheons to that of polytheistic Egypt or brought their own mysterious monotheism with them, thought to be centred around a providential, life-giving sky god who was possibly called Waaq. In addition, Alexandria was also home to the largest urban population of Jews in the ancient world, while the nearby Natron Valley played host to the Desert Fathers and Mothers, the ascetic hermits that had a profound impact on the development of Christian theology, who had gathered from across Christendom to devote themselves to honing the tenets of their new faith out in the wilderness. Centuries before the prophet Mohammed's inception of Islam, Arab visitors from the peninsula just across the Red Sea may have followed interesting flavours of early Judeo-Christian faiths, Zoroastrianism, Manichaeism, or an indigenous polytheistic religion revolving around creator god Allah, three goddesses called al-Lāt, al-'Uzzā, and Manāt, and a tangled, localised folkloric pantheon involving other minor deities, djinn, and the circumnambulation of sacred stones. The vast Roman Empire, meanwhile, with its well-travelled, socially mobile, ethnically diverse legions and civil servants, had a taste for mystery cults, which it liberally imported to and exported from all of the territories it conquered. One of the Greek Magical Papyri is referred to as the Mithras Liturgy, after the mystery cult devoted to the Zoroastrian divinity Mithra that spread extensively throughout the Roman Empire, although the text itself seems to have more in common with the later alchemical ideal of the magnum opus than it does with what little we know of Romano-Persian Mithraic rites. The Greeks had a vast range of intellectual and proto-scientific philosophies as well as a polytheistic religion that had been deeply syncretised with both the Roman and Egyptian pantheons, while the native Egyptian religion itself was among the oldest and most complex faiths the world had ever seen, and its practitioners were famed for the quality of their scholarship on a variety of subjects, notably cosmology, medicine, and the manufacture of paints, dyes, cosmetics, scents and other potions. In the capital of Graeco-Roman Egypt, elements of all of these diverse traditions - some ancient, some new - came together like the result of a successful alchemical experiment to form something unique: a clever, questioning, technically adept society that intrigued and attracted other curious minds from a broad spectrum of faiths and cultures.

The city wasn't a multicultural utopia. There were plenty of ideological clashes between religious groups - gentiles versus Jews was a frequent flashpoint, as was pagans versus Christians: in the 2nd century CE there had been pogroms against the Jewish community, and answering riots as they rebelled against the persecutions of the Kitos War; Christians were frequently oppressed and even martyred by both Greek and Roman authorities. Nonetheless, despite the tensions that occasionally flared, ancient Alexandria was a beacon, just like the one that graced the roof of the iconic lighthouse that towered over its harbour from the tiny island of Pharos, one of the Seven Wonders of the Ancient World. The Pharos lighthouse became an enduring symbol of the city that was one of the ancient world's leading lights in intellectual and spiritual development. Alexandria was a place where Greek philosophy could come together with the long-studied theologies of Judaism and its then more mutable, much younger cousin, Christianity, as well as the diverse beliefs of other residents of and visitors to the city. A century or so before the confluence of these and many other ideas would come together to create the beginnings of alchemy, they had already begun to cross-pollinate. The first result of this seemingly unlikely combination is now known as Gnosticism.

Academic debate still rages as to whether Gnosticism was a religion in its own right or an inter-faith intellectual movement, although it's likely that the original Gnostics would probably be somewhat bemused by the fact that we make a distinction

"APOLLONIUS OF RHODES, AUTHOR OF THE ARGONAUTICA, WAS ONE OF ITS HEAD LIBRARIANS"

The Library of Alexandria was once the largest in the ancient world and was packed with texts on philosophy, theology mathematics, natural history, alchemy and much more

Alexandria's multicultural khēmia artisans were masters of glassmaking, metalwork, colour, paint and dye technologies, as shown by this miniature glass tile

The Library of Alexandria experienced several attacks: accidental burning by Julius Caesar in 48 BCE when he fired his own ships, an attack by the Roman Emperor Aurelian in 272 CE, and finally Emperor Diocletian's siege of Alexandria in 297 CE

between the two concepts. Their complex worldview took and combined theological ideas from Judaism and Christianity with Greek philosophy and the cosmological elements of Persian Zoroastrianism. While the origins of Gnosticism are murky, it's safe to say that we have no records of it that predate Christianity, and we know that Alexandria was very important to its early development. Early Church Father Clement of Alexandria, an ethnic Greek who converted to Christianity, was certainly familiar with its synthesis of Greek philosophy, Judeo-Christian theology and Eastern mysticism, although it's debatable whether he was a practising Gnostic or not. Like many intellectuals of the time, Clement made a distinction between his public works and other more secret records that he kept, and it's these that contain tantalising glimpses of Gnostic thought. Throughout much of Christendom, Gnosticism was considered a heresy - the first to emerge within Christian theology. The Alexandria-educated 2nd century Egyptian theologian Valentinus broke with the Catholic church when it passed him over for a promotion and went on to embrace the philosophy to such an extent that a branch of Gnosticism - Valentinianism - is named after him. 'Valentinian' was possibly the first term that was used to refer to Gnostics - the original followers of the idea used 'gnostikos' (Greek for 'having wisdom') to describe their philosophy, not themselves. 'Gnosticism' and its associated labels is actually an 18th century construct.

Movements like Gnosticism found such a welcoming home in Alexandria because what we would term its middle class - scholars, artisans, merchants, civil functionaries and the like - had a vibrant, multi-ethnic, knowledge-seeking mindset. At the height of its powers, Alexandria's famed library was the largest in the ancient world, part of a vast complex called the Mouseion, an institute somewhat like a modern-day research university, dedicated

EARLY ALCHEMISTS

THE EARLIEST ALCHEMISTS ARE MORE LEGEND THAN HISTORICAL FIGURE

The earliest figures in Alexandrian alchemy are semi-legendary. In a world in which similar-sounding words across different languages were enough to prove a deep connection, an early alchemist, and one who apocryphally was thought to have given his name to the art, is Chymes, known in Arabic manuscripts as 'Kimas' or 'Shimas'. In some Alexandrian Graeco-Jewish traditions he was associated with Ham, the son of the biblical Noah, who was thought to have settled in Egypt after the Flood. Another early alchemist with an Abrahamic name is the mysterious Moses of Alexandria, who is associated with the prophet of the same name - in ancient Alexandria the biblical Moses was thought to have invented the arts and sciences. Then there were the mysterious figures who wrote under the pen names of Greek philosophers - Pseudo-Democritus, Pseudo-Aristotle and Pseudo-Plato. We know about many of these figures from the writings of Zosimus of Panopolis. He praised many of them, but wrote to his sister Theosebeia, also an alchemist, not to continue corresponding with one Paphnutia the Virgin, saying she was under-educated and incorrect. It's thought that the earliest historically provable practicing alchemist was Mary the Jewess.

A piece of syncretic Alexandrian folklore made the biblical Ham, son of Noah, into Chymes, an early alchemist after whom it was claimed alchemy was named

to the Nine Muses of Greek myth and, at its peak, packed with over 1000 well-paid, freethinking scholars busily collecting knowledge from throughout the known world and translating it for the benefit of its large membership. Apollonius of Rhodes, author of the epic poem the *Argonautica*, was one of its head librarians; during his tenure, the famed mathematician and inventor Archimedes visited. His followers claimed that observing the Nile allowed him to come up with his famous water pumping device, the Archimedes screw, but it's more likely that the Sicilian-Greek engineer got the idea from records of ancient Egyptian, Sumerian or Persian irrigation technologies that he found in the extensive scroll collection of the Great Library. By the time of the Roman occupation, however, membership of the Great Library was much fallen off, and it met its sad end under the auspices of either the Emperor Aurelian in 272 CE or the Emperor Diocletian in 297 CE. Diocletian is a likely candidate: the Croatian-born scribe's son was a boot-strapped army commander who rose to the purple to become a Rome-first traditionalist. He persecuted dissenters and minorities throughout the Empire during his rule and, after a visit to Alexandria, issued decrees that condemned "the ancient writings of the Egyptians, which treat of the khēmia transmutation of gold and silver" and encouraged the burning of alchemical texts. It's thought that this is when the original Alexandrian version of fabled text the *Hermetica* was lost.

As the namesake of Alexander the Great, the home of the ancient world's greatest library and one of its Seven Wonders, and the place where alchemy was invented, Alexandria was idealised by the West in the Middle Ages

Alchemy - or khēmia, as it was then known - is thought to have begun around a century before Diocletian's ban. Its earliest known figures include Mary the Jewess and Cleopatra the Alchemist, Zosimus of Panopolis and master of poisons Agathodaemon, who used the local natron salts in combination with arsenic-bearing minerals to produce the clear, deadly arsenic trioxide. Much of their early knowledge is likely to have come from metallurgy: arsenic, for example, can be used to harden bronze, and this is how its poisonous nature was discovered. But while alchemy has its founding myth of Hermes Trismegistus, the actual historical record of its inception is somewhat sparse. These early practitioners are recorded as among the first to have developed scientific instruments like the bain-marie (Mary) and the alembic (Cleopatra) as well as some of the first alchemical processes. But could the proto-chemists of ancient Alexandria really transmute base metals into noble ones? They could certainly appear to do so. An Alexandrian record now known as the Leiden Papyrus preserves the formula for an important recipe known as 'water of sulphur'. A carefully measured combination of calcium oxide, sulphur and acid (the recipe suggests vinegar or urine), when correctly prepared and applied it can turn silver to the rich, gleaming colour of gold. Science historian Lawrence M Principe, a practicing chemist as well as a historian of alchemy, has replicated the substance and found it to work. These kinds of transformations are what the early chemists of Alexandria specialised in: they did a roaring trade in producing fake metals and gems for the Egyptian city's colourful, well-dressed clientele. For Diocletian, struggling with Empire-wide economic disasters brought about by inflation, forgery, coin clipping, and exchange rate inconsistencies that not even a new tax levy could fix, the idea that the alchemists of Alexandria could pass silver off as gold was terrifying. Although ancient Rome has a reputation for looking askance at any kind of esoteric practice that it hadn't accepted as its own, Diocletian's proscriptions weren't so much about a fear of non-Roman sorcery as they were about the very real financial dangers of a new and disruptive technology. Alexandria's alchemists were nothing less than a threat to the fiscal stability of the entire Roman Empire, an empire that within Diocletian's reign had revealed itself to be too large and unwieldy for one ruler to manage and had fractured into the First Tetrarchy of four regional leaders. Diocletian's attack on the alchemists of Alexandria was nothing new - he was a keen persecutor of Manicheans and Christians too - but in attacking the alchemists, and more importantly, the records of their knowledge and works, Diocletian effectively tore the heart out of the prosperous city that early alchemy called home. Alexandria had bounced back from the brink before - it had lost its founding father, misplaced his tomb, experienced disease, disaster and death; it had even coped with a predecessor of Diocletian's massacring over 20,000 people, but one thing it couldn't handle was burning books. Diocletian's edict and the loss of its library sucked the life out of the city. It limped along until a tsunami driven by the 365 CE Crete earthquake devastated it. But the seeds of alchemy had been left in the wreckage, and they would eventually be rediscovered by the city's Arabic conquerors, who would continue the alchemists' research until it spread and eventually became modern science. Perhaps Aristander was right when he told Alexander that his Egyptian city would one day nourish the world.

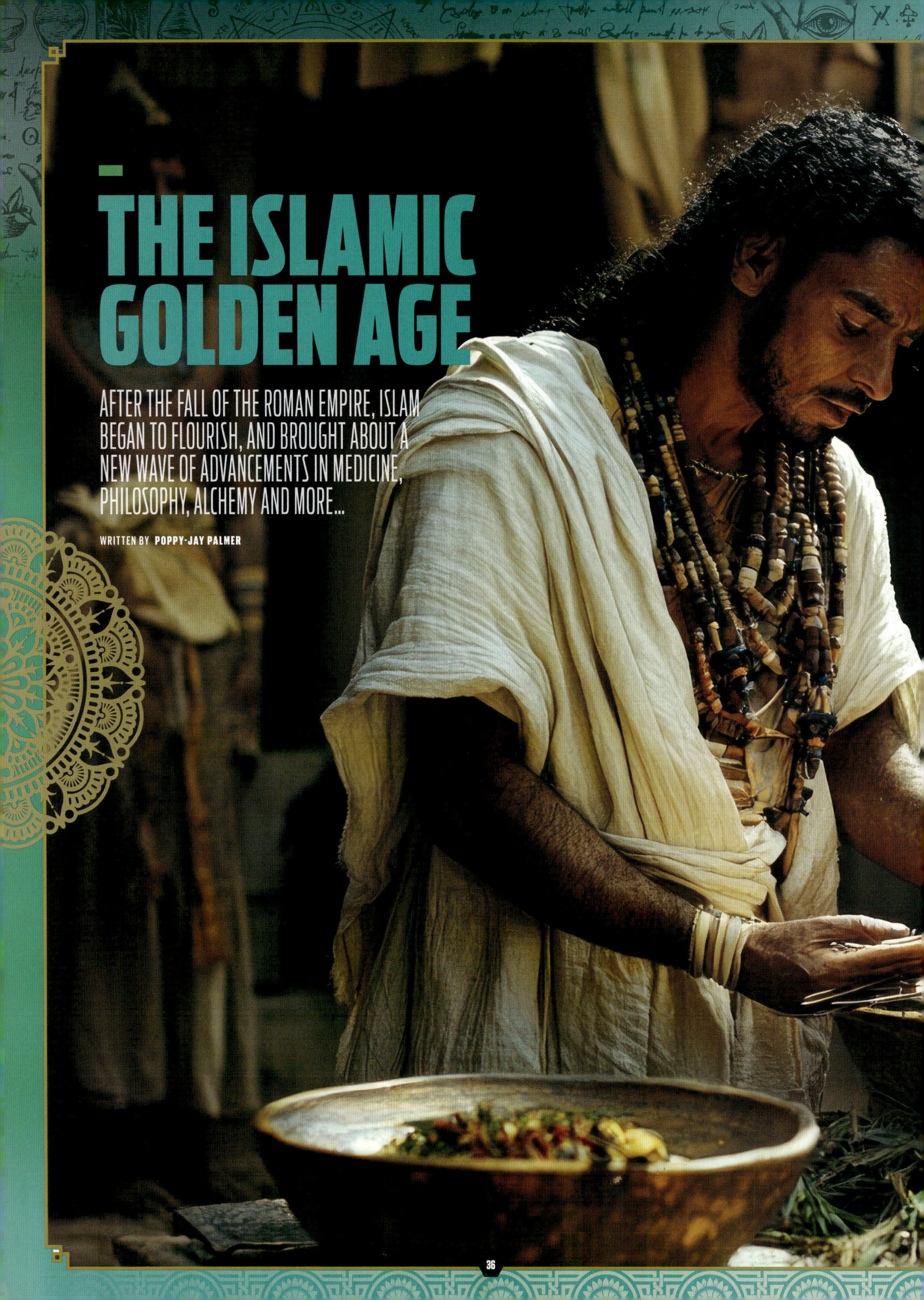

THE ISLAMIC GOLDEN AGE

AFTER THE FALL OF THE ROMAN EMPIRE, ISLAM BEGAN TO FLOURISH, AND BROUGHT ABOUT A NEW WAVE OF ADVANCEMENTS IN MEDICINE, PHILOSOPHY, ALCHEMY AND MORE...

WRITTEN BY **POPPY-JAY PALMER**

From the 8th to the 14th century, Islam flourished. The growth of the religion spurred on a period of cultural, economic and scientific advancement like no other. Following the death of Muhammad, the caliphs – the new Arab leaders – built and established a new city, Baghdad, as the heart of the Abbasid Caliphate. Conveniently located between Europe and Asia, Baghdad was an integral area for trade and the exchange of ideas. Over time it transformed into a hub of learning and commerce and, for a while, it became an unrivalled centre of science, medicine, education and philosophy. The period welcomed in what is now referred to as the Islamic Golden Age.

Knowledge was regularly shared at the famous House of Wisdom, or the Grand Library of Baghdad, where scholars from far and wide and with different cultural backgrounds gathered to translate all of the world's classical knowledge into the Arabic language. As insight into topics like art and culture, healthcare, law, theology, engineering and natural sciences expanded, so did the understanding of alchemy.

After the fall of the Western Roman Empire, the study of alchemical development moved to the Caliphate and the Islamic civilisation, which became one of the world's leading cultures when it came to both traditional alchemy and early practical chemistry. Even the word alchemy was originally derived from the Arabic word al-kīmiyā. Some historians also believe that it came from the Egyptian word kemi, meaning black. Much more is known about ancient Islamic alchemy that that of other cultures owing to the fact that Islamic alchemy was far better documented, with most of the earlier writings that have been passed down through the years being preserved as Arabic translations of the original Greek, Roman and Egyptian.

Ancient Islamic chemistry – or Arabic chemistry, as it is also often called – covers a whole matter of topics. The late AI Sabra, a professor of the history of science who specialised in the history of optics and science in medieval Islam, narrowed down the definition. In his article Situating Arabic Science: Location versus Essence, he described Arabic (or Islamic) science as a term describing the scientific research and activities of people who lived in a region that roughly extended chronologically from the eighth century CE to the beginning of the modern era, and geographically from the

During the Islamic Golden Age, the city of Baghdad became a hub of learning and commerce and changed Middle Eastern culture forever

Iberian Peninsula and North Africa to the Indus Valley and from southern Arabia to the Caspian Sea, or the region covered by what has been described as Islamic civilisation for most of that period. The research and findings of Islamic chemistry were mostly expressed in the Arabic language. Ancient Islamic alchemy, on the other hand, refers to a very particular subset of Islamic chemistry: the search for metallic transmutation.

Though different cultures had different approaches when it came to the sciences during the Middle Ages, there was often overlap between that of the Arabic areas and that of the Western hemisphere. Cultural, religious and scientific diffusion of information between Eastern and Western societies began with the conquests of Alexander the Great in the 3rd century BCE, with greater communication between the two being allowed. Thousands of years later, those Eastern territories that had been conquered, like Iran and Iraq, became the centre of religious movements, including Christianity, Manicheism and Zoroastrianism, all of which involve sacred texts as a basis and encouraged literacy and the spread of ideas, and the Qur'an, Islam's holy book, became an important source of theology, morality, law and cosmology. Following Muhammed's death, Islam spread throughout the Arabian Peninsula, Persia, Syria, Egypt, Israel and Byzantium by way of military conquest. With it went a keenness for knowledge and scholarship, and the concept of sciences like alchemy. Islamic alchemy was studied as a subset of science, but it still held onto the mystical and religious aspects that set Eastern alchemy apart from the brand that was studied in the Western hemisphere, which predominantly held Christian ideals. Just as the work of the likes of the so-called Hermes Trismegistus is studied today in order to better understand the workings of ancient Western alchemy, Islamic alchemy has many rockstars of its own. Jābir ibn Hayyān, one of the most famous Islamic alchemists, covered a vast range of topics – including cosmology, philosophy, astrology and more – in his written works known as the Jabirian Corpus. Khalid ibn Yazid, also known as King Calid, was the author of the first alchemical work translated from Arabic to Latin.

Perhaps not as well known as Jābir or King Calid but just as notable was Dhūl-Nūn al-Misri, an early Egyptian mystic and ascetic who studied alchemy, medicine and Greek philosophy in his early life. Both during his lifetime and after, Dhūl-Nūn's work and legendary wisdom have caused him to be celebrated by Islamic thinkers and considered one of the greatest saints of the early era of Sufism. As a legendary thaumaturge, he was supposed to have known the secret of the Egyptian hieroglyphs. None of his written works have survived the centuries but his sayings and poems, both rich in mystical imagery, endured in the oral

Islamic scholars translated as much classical knowledge from Latin to Arabic as possible, meaning historians have been able to learn from the records

Rhazes is well-known for discovering numerous compounds and chemicals, including alcohol and sulphuric acid, through his work in alchemy and chemistry

Jābir ibn Hayyān and Khalid ibn Yazid led the way when it came to alchemy and its study during the Islamic Golden Age

Muhammed ibn Umail al-Tamini viewed alchemy symbolically, which became an influential idea

tradition and brought him even more fame and appreciation over the years. Although much of Dhūl-Nūn's legacy revolves around the mystical side of him, he was also a well known Hermetic alchemist and is often associated with Jābir ibn Hayyān. Another influential Islamic alchemist was Muhammad ibn Zakariya al-Razi, also known by his Latinised name Rhazes. He was also a polymath, physician, philosopher and an important figure in the history of medicine. Through his work, Rhazes made a number of fundamental contributions to a number of different fields: he wrote a pioneering book on smallpox and measles and their characteristics, carried out integral work on the understanding of ophthalmology, or the diagnosis and treatment of eye disorders, and was even the first person to recognise the reaction of the pupil to light. Through translation, this medical work was passed around European practitioners and profoundly influenced medical education in the West, with some of his books making it to the medical curriculum in the first Western universities.

However, it was his work with compounds and chemicals that caught the attention of alchemists all over the world, and many of the chemical instruments he developed are still used today. Within alchemic and chemistry circles, Rhazes is well-known for discovering numerous previously undiscovered compounds and chemicals, including alcohol, which he obtained through perfecting methods of distillation, and sulphuric acid. He also strongly believed in the possibility of transmutation of lesser metals into silver and gold, a theory that was attested after his death by Arab scholar Abū al-Faraj Muhammad ibn Ishāq al-Nadīm's chapter on alchemy in his book *Kitāb al-Fihrist*. Rhazes was so renowned in the field that some of his contemporaries even believed that the alchemist had obtained the secret of transforming iron and copper into gold. His studies can be better understood through reading his many texts on alchemy, most of which are written in Persian.

The work of 10th-century alchemist Muhammed ibn Umail al-Tamini gives insight into a different side of alchemy that was alive and well during the Islamic Golden Age. Very little is known about his life, as he famously lived in seclusion, but his writings suggest that he was born in Spain to Arabic parents before living and working in Egypt. As a mystical and symbolic alchemist, he was known for rejecting alchemists who took their subject too literally. Where some of his peers studied the sciences of metallurgy and chemistry, Muhammed ibn Umail al-Tamini preferred to focus on the symbolic meaning of alchemy, which he believed was tragically overlooked. In his work *Book of the Explanation of the Symbols*, he emphasised that the sages spoke in symbols and explained the study of alchemy as an allegory for something more complex. Though he was devoted to Greek alchemy, he wrote as a Muslim and frequently mentioned his religion in his work and even quotes verses from the Quran. As well as studying the symbols of alchemy, Muhammed ibn Umail al-Tamini also presented himself as an interpreter, and set his treatise *Silvery Water* in the Egyptian temple known as the Prison of Yasuf, where the Jewish, Christian and Muslim figure Joseph famously interpreted the dreams of the Pharaoh.

AGE OF GOLD

'Golden Age' might seem like quite a grand way of describing Islamic culture at the time of the Middle Ages, but it's also rather fitting. The term sits in contrasts to the European Dark Ages, the historical period that occurred in the Western world shortly after the fall of the Roman Empire and saw a huge demographic, cultural and economic decline.

The term 'Islamic Golden Age' first started being used in 19th century literature about Islamic history in the context of Orientalism, the western aesthetic fashion of imitating or depicting aspects of the Eastern world. The term was notably used by the author of *A Handbook for Travellers in Syria and Palestine*, published in 1868, which said that Damascus' most beautiful mosques were "like Mohammedanism itself, rapidly decaying," and relics of "the golden age of Islam."

As well as describing the culture of the time, "Islamic Golden Age" was also used to comment on military achievement. The timespan differs depending on the context. Some historians extend the Golden Age to the duration of the Caliphate, while others have it end after a few decades of Rashidun conquests with the death of Umar and the First Fitna. Since the second half of the 20th century, the term has mostly been used to describe the cultural side of things.

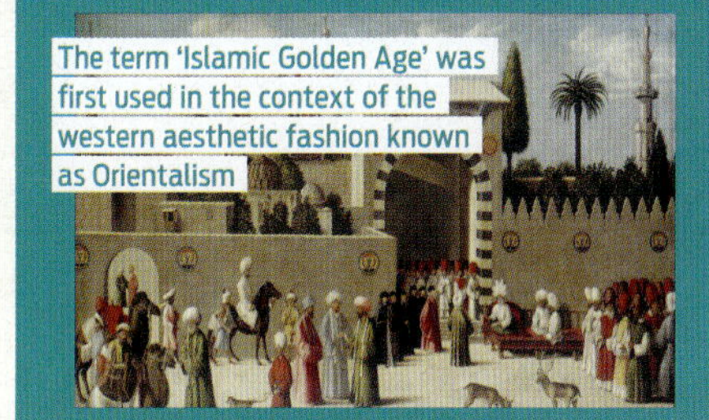
The term 'Islamic Golden Age' was first used in the context of the western aesthetic fashion known as Orientalism

Images source: Getty, Wiki

Scholars from across Europe travelled to Spain to study under Arabic teachers and returned with knowledge of many lost arts – including alchemy

ALCHEMY ENTERS EUROPE

ALCHEMY WAS A LOST ART IN LATIN EUROPE UNTIL BOLD TRANSLATORS OF ARABIC WORKS BEGAN ITS REINTRODUCTION. ONCE AGAIN, IT GRABBED THE SCHOLARLY IMAGINATION

WRITTEN BY **BEN GAZUR**

On 11 February 1144, a monk laid down his quill on a work that would resonate throughout Europe for centuries. Robert of Chester had just completed his translation of an Arabic text that first introduced the word alchemy to Latin Europe. We know the exact date because he tells us in his preface to *Liber de compositione alchimiae* - 'The Book of the Composition of Alchemy'. He undertook this work "because our Latin world does not yet know what alchemy is." Very soon everyone would have heard of it.

Out of Iberia

Often Europe of the Middle Ages is considered as a monolithically Catholic Christian area. This is far from the truth. The Iberian Peninsula, including much of what we now call Spain, was under the control of Muslim rulers. This area, known as Al-Andalus, came into being in 711 when a force under the Ummayad caliphs invaded from North Africa.

Within 40 years most of the peninsula was under Ummayad control and the kingdom of the Franks was being threatened by their forces. Only with great difficulty was the expansion of the Muslim forces stopped at the Pyrenees. Portions of Iberia would remain under Muslim control until the surrender of the Kingdom of Grenada to Isabella of Castile in 1492. In the intervening years, however, the regions of Muslim Iberia would be among the most advanced in all of Europe. The mixing of Arabic and Latin influences in Al-Andalus allowed for an unprecedented exchange of knowledge between the Islamic and Christian worlds. Al-Andalus became the place to go for European scholars looking to learn all that could be learned at the time. Arabic scholars had one great advantage over their European rivals at this time - they had access to the great works of Greek science and literature that had long been thought lost in the West.

Greek had once been the lingua franca of the ancient world and many learned treatises were written in it. As Latin overtook it in Western Europe as the language of education, many works and writers were forgotten. After the Arab conquests of the 7th century, translations of these texts in Arabic became freely available. Now if a Latin speaking scholar wished to know what Galen or Aristotle wrote they had to learn to speak and read Arabic.

This is exactly what Robert of Chester did. In the 1140s he travelled from England to Al-Andalus and translated several important works from Arabic to Latin. As well as his alchemical translations he produced a Latin version of Al-Khwārizmī's book about mathematics, which introduced the idea of algebra to the West.

Robert Who?

Robert of Chester was far from the only

Images source: Adobe Stock, Getty

As well as introducing the word alchemy into Latin, Robert of Chester described what would become known as the Philosopher's Stone

The alchemical text translated by Robert of Chester describes Prince Khalid receiving the wisdom of a learned hermit

scholar to beat a path to Al-Andalus to learn from Arabic sources. Pope Sylvester II had spent time there during his education, which led to suspicions that he had been taught dark and heathen magic such as the creation of a brazen head that could answer any question put to it. Despite the dark mutterings that might have attached to some of the wisdom learned in Al-Andalus, many others saw training there as a stepping stone to academic advancement. So many went that there is some confusion as to who deserves the credit for some of the translations of the time.

Robert of Chester, Robertus Castrensus, is usually said to be the author of the Latin translations mentioned above, but there is some confusion between Robert of Chester and one Robert of Ketton. Like Robert of Chester, Robert of Ketton was an Englishman who went to Al-Andalus in the 1140s to translate Arabic texts into Latin. Because of their similar names and the scarcity of records in the past, the two have often been conflated into a single individual.

While we know little about Robert of Chester except what he himself put into his works, there is rather more on record about Robert of Ketton. Robert of Ketton undertook the first translation of the Quran into Latin, though he called it "The Law of the False Prophet Muhammed". His later career as a clergyman and royal advisor is well documented. Despite some similarities, most historians are convinced that the two Roberts really were different individuals doing similar work at the same time.

"MORIENUS USES THE METAPHOR OF THE TAILOR TO EXPLAIN HOW MANY THINGS CAN BE MADE OUT OF A SINGLE PRIME SUBSTANCE"

The Legend of Khalid

Robert of Chester's alchemy book *Liber de compositione alchimiae* was based on an Arabic text called *Masā il Khālid li-Maryānus al-rāhib* - 'Questions of Khalid to the Monk Maryanus'. This work, likely from the 9th or 10th centuries, purports to retell the tale of the meeting between Prince Khalid and a hermit skilled in alchemy, Maryanus. Morienus, as his name is translated, proceeds to impress his royal pupil with his wit and wisdom. In Robert of Chester's retelling of events Morienus is a pious Christian from whom the Muslim Khalid learns, while in the Arabic original he is simply described as a "Greek" instead.

Morienus cautioned Khalid that "The ancients, however, did not refer to the matters pertaining to this science by their proper names, speaking instead, as we truly know, in circumlocutions, in order to confute fools in their evil intentions." This is a point on which the royal interlocutor is not satisfied and asks for plain language to be used.

The royal Khalid sometimes becomes impatient with the riddling nature of some of the answers. When his request to know where the magical Philosopher's Stone is, and how it can be inside everyone, draws an uninformative answer, however, Morienus simply shrugs and replies that he has already answered the question.

Morienus uses the metaphor of the tailor to explain how many things can be made out of a single prime substance. A tailor may take a piece of cloth, cut it in a number of ways, and make many different garments. In the same way the alchemist takes matter, pulls it apart, and reforms it to his desired end. Morienus does indeed use some of the same language that will be repeated for centuries in European alchemy.

When pressed he explains his obtuse allegorical meaning in chemical terms. "The green lion is glass... and the stinking Earth is stinking sulphur... But red fume is red Auripigmentum, also white fume is Argent Vive, and citrine fume is citrine sulphur... Behold now I have expressed to you the names of kinds, of which three will suffice you for the whole Magistery: that is, white fume, the green lion, and stinking water."

Alchemy finds a home

Robert of Chester's translation of Arabic works into Latin would not be the last. When Robert of Chester used the word alchemia he was not discussing the wider subject as we would know it but rather just what would become the Philosopher's Stone. Alchemia was a "substance... [that] naturally converts substances into better ones." Only with later translations of Arabic texts would the word alchemy come to take on its full meaning. It should not be supposed however that as soon as Robert of Chester set down his quill in 1144 that the whole of Europe became enamoured with alchemy. The spread of this secret knowledge took quite some time. When Roger Bacon came to advise the Pope on matters of education over a century later, he felt it necessary to explain just what alchemy was to the pontiff.

"There is another science which is about the generation of things from the elements, and from all inanimate things,... of which we have nothing in the books of Aristotle; nor do natural philosophers know of these things, nor the whole Latin crowd of Latin writers. And since this science is not known to the generality of students, it necessarily follows that they are ignorant of all natural things that follow there from... And this science is called theoretical alchemy, which theorises about all inanimate things and about the generation of things from the elements."

After the first translations were made out of Arabic, however, European alchemists soon began to create their own works rather than rely on the borrowings from others.

They used the mystical and metaphorical language of the historical alchemists they found in the translations, but used them for their own ends. Robert of Chester had introduced alchemy to a fertile field of medieval imagination, which was already enamoured with allegory and symbols.

While European alchemists were creating their own theories, however, they did acknowledge the founders of their art. Many published their works not under their own names but under those of Arab thinkers such as Jabir, al-Razi, and ibn-Sina.

ROBERT MAKES A BOOB

Translation can be tricky work. The subtle meanings of a word can be lost when they are transferred from one language to an altogether different one. Sometimes the mistakes made in translation can come from a complete blunder, but last for centuries.

When Robert of Chester came to translate Al-Khwārizmī's book on algebra he came across a word for a trigonometric function of an angle. This text used the Arabic word jb to describe the ratio of the length of the side of a right-angled triangle that is opposite an angle to the length of the longest side. Today we would call this function sine, but that is only because Robert of Chester made a mistake in his translation.

The text he was translating had imported a Sanskrit word, jya-ardha (shortened to jya) into the Arabic work without translating it, merely spelling it phonetically as jiba. Since Arabic is written without vowels, this appeared as jb. Robert took this word to be jaib, meaning breast or bosom, and translated it literally into Latin as sinus. Thus, whenever a sine function is being used, you can think of Robert of Chester and the time he failed to get fully abreast of his subject.

Even the most careful scribes could make mistakes when faced with a text in a different language on a technical subject

The alchemical metaphors and symbolism in Robert of Chester's translation shaped how alchemy was thought about for centuries

The inability to read Greek texts, such as Euclid's Elements, meant that knowledge of them often came via Arabic translations

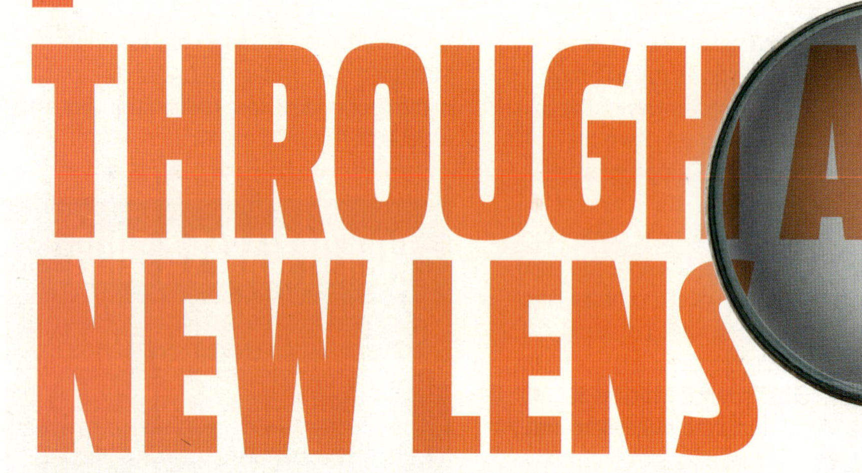

THROUGH A NEW LENS

THE DEVELOPMENT OF OPTICS IMPROVED WAYS OF SEEING IN MORE WAYS THAN SIMPLY CORRECTING POOR VISION...

WRITTEN BY EDOARDO ALBERT

The writer typing these words is wearing glasses. As such, he belongs to a fortunate portion of humanity: those born after the invention of spectacles. For most of history, and for all of prehistory, there was nothing people could do about the weakening of sight. Cicero, the Roman orator, was rich enough to have slaves read out loud to him, but most people had to make do with blurred vision.

Then, around 1290, an unknown craftsman in Italy, probably living in Pisa or Florence, discovered how to make spectacles. We know this because a few years later, on 23 February 1306, Giordano da Pisa, a Dominican friar, preached a sermon in which he said, ""It is not yet twenty years since there was found the art of making eyeglasses, which make for good vision." Giordano went on to say that he had met and talked with the inventor.

However, this unnamed inventor, in an early attempt at covering up trade secrets, refused to tell anyone else his methods. But Dominican friars were the cleverest and best educated men of the Middle Ages. One of Giordano's brethren, a friar named Alessandro della Spina, worked out how to make spectacles himself and, unlike the original manufacturer, cheerfully shared the knowledge of how to do so with anyone who wanted to learn.

Of course, to make good quality spectacles you need good quality glass. Fortunately, just on the other side of the Italian peninsula there was the Venetian island of Murano, already famous for the quality of its glassware. Spectacle manufacturing spread rapidly to Venice: by 1320 there was a Venetian guild of spectacle makers who rapidly began making glasses in very large quantities. The trade spread quickly and widely: first through Italy and Europe, and then further afield, with shipments of tens of thousands of spectacles sent to the Middle East.

These early spectacles all had convex lenses, which meant that they corrected for long sight, but they did not help short-sighted people. They had to wait until the late 15th century for spectacles to be fitted with concave lenses. We know this, extraordinarily, because around 1518 Raphael painted Pope Leo X holding a reading lens and the painting is detailed enough to show that the lens is magnifying what is underneath it.

The invention of spectacles was a happy one for anyone without perfect 20:20 vision. But it also ushered in the ability to magnify that which was small and bring closer that which was far away. Soon, visionaries would turn these new eyes to the starry heavens above and to the microscopic world below.

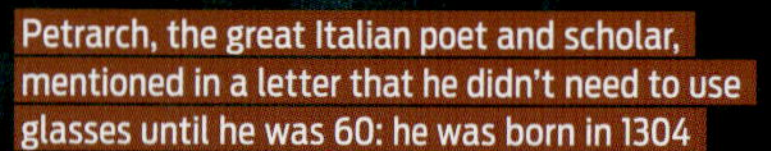

Petrarch, the great Italian poet and scholar, mentioned in a letter that he didn't need to use glasses until he was 60: he was born in 1304

The earliest painting of a man wearing spectacles: Tommaso da Modena's 1352 portrait of Cardinal Hugh of Saint-Cher

Raphael's stunningly naturalistic portrait of Pope Leo X shows him holding a concave reading lens in his left hand above the Bible for which he was using it to read

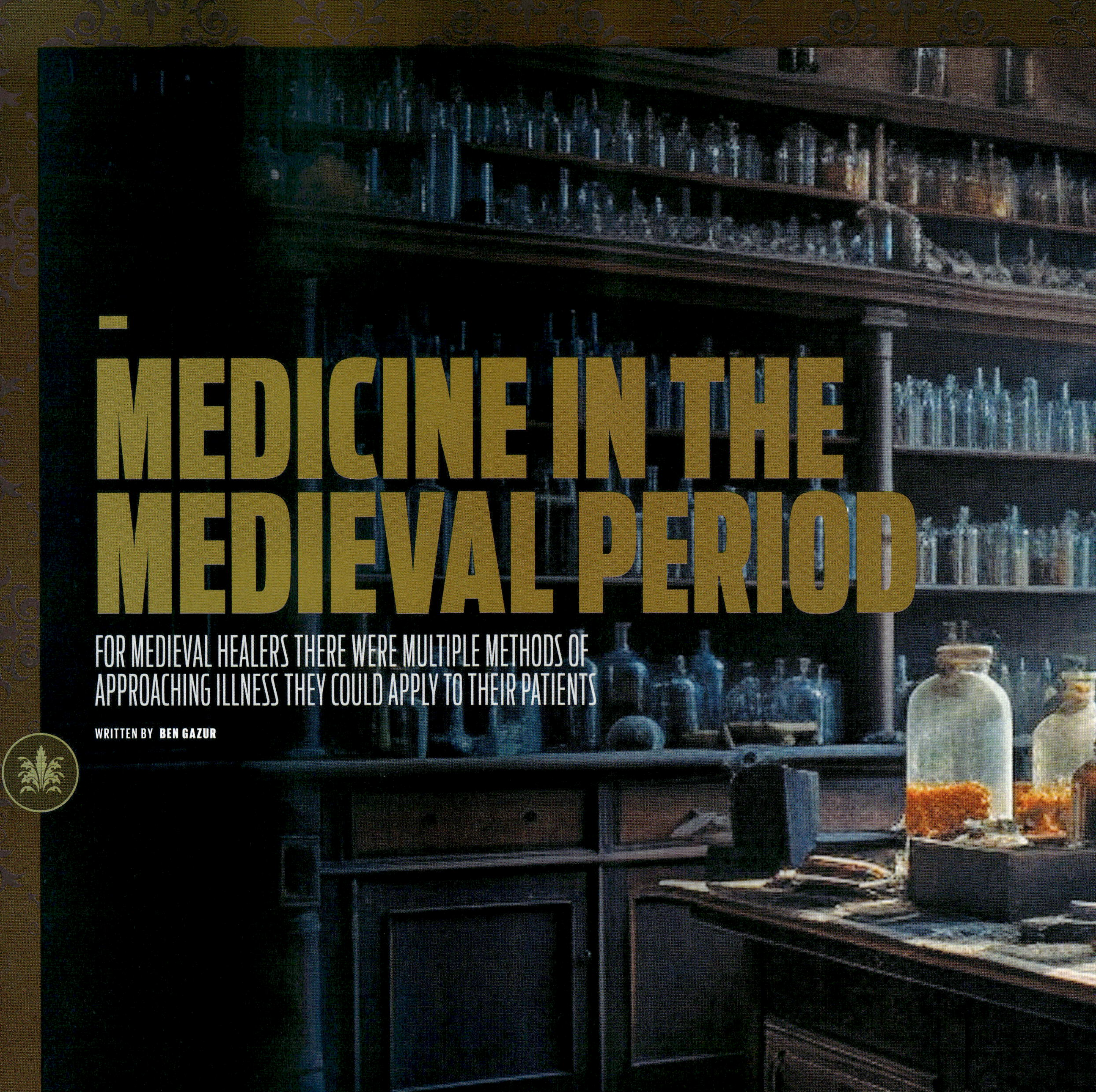

MEDICINE IN THE MEDIEVAL PERIOD

FOR MEDIEVAL HEALERS THERE WERE MULTIPLE METHODS OF APPROACHING ILLNESS THEY COULD APPLY TO THEIR PATIENTS

WRITTEN BY **BEN GAZUR**

With all the wonders of modern medicine it is easy to look at medical practitioners of the medieval period as deluded quacks who relied on credulity and superstition. While it is true that many people relied on treatments which would have done little to cure illnesses there was a great deal of study and theory which underpinned how healers operated. Medieval doctors drew on ancient sources, the most modern thought, and empirical experience to create a remedy - though some of these attempts strike us as bizarre.

The medieval period, or Middle Ages, lasted from the fall of the Western Roman Empire in the 5th century until around the 15th century. This millennium of European history includes a huge variation in lifestyle, nations, and knowledge. It can be difficult to speak of the medieval period as a single unit. The practice of medicine changed dramatically over this time.

Ancient Origins

Medicine in the medieval era was heavily influenced by authorities who worked in ancient Greece and Rome. The Hippocratic texts, the works of Galen, and those of Pedanius Dioscorides were copied extensively in Western Europe and formed a nucleus of study for doctors. These texts gave a foundation of diagnosis, anatomy, surgery, and treatment.

After the collapse of the Roman Empire in the West there was a strong tendency to treat the fallen empire as a lost golden age. Sources of knowledge which came from ancient days were highly regarded, in many cases more so than practical experience. Antiquity lent authority. Doctors who could point to a Roman or Greek work as the basis

for their treatments was more likely to be trusted by potential patients.

There were reasons to rely on ancient texts. The study of human anatomy in the medieval period was limited because of Church bans on the dissection of dead bodies. Most doctors would only have known anything about the internal organs of a person from reading about them in the works of ancient authors. There were downsides to trusting these texts. Galen seems never to have dissected a human and based his work on cutting up apes and believing the human body was exactly the same. Healers who had experience of battle probably had better anatomical knowledge.

Perhaps the most influential ancient theory of medicine was the Hippocratic belief in the four humours: blood, phlegm, black bile, and yellow bile. Disorders could, it was thought, be explained by levels of these substances being out of balance in the body. For centuries doctors sought to cure patients by removing blood or giving medicines to provoke vomiting to redistribute and balance the humours.

The textbooks of medicine used for most of the Middle Ages had to be copied

Medieval doctors considered a number of factors about their patients before deciding on the best course of treatment

Images source: Alamy, Adobe Stock

repeatedly by hand if they were not to be lost forever. Most scribes were trained clerics so many works were to be found in the libraries of monasteries. These became the nucleus of many of Europe's great schools of medicine.

Christian Influence

It is impossible to comprehend medieval Europe without an appreciation of the power and influence of the Catholic Church. This extended to acceptable medical training. Ancient writers were tolerated because their works did not contradict Catholic theology. Many in the Church might think that illness was caused by God as a punishment for sin but they also allowed doctors to do their best to help. The early Church Father Origen had declared that doctors were appointed to minister to the sick by using the healing substances God in his wisdom had placed on the Earth.

Since many of the finest texts were found in monasteries it was natural that many of the best trained healers were found in monastic establishments. There was a theological reason too as to why many monasteries served as hospitals. The Holy Rule of St Benedict was followed in many abbeys and this told monks 'Before and above all things, care must be taken of the sick, that they be served in very truth as Christ is served; because He hath said, "I was sick and you visited Me"'. Special cells were set apart and the ill given dispensation to eat meat if it would help their recovery. Members of a religious order could become renowned for their healing skill, such as German abbess Hildegarde of Bingen.

Monasteries and nunneries grew much of their own food and so would have an extensive garden. Some of the space of the garden was given over medicinal plants such as herbs that could be used to treat illnesses. Aromatic herbs like rosemary and mint found a place both at the dining table and the sickbed.

> "MANY OF THE BEST TRAINED HEALERS WERE FOUND IN MONASTIC ESTABLISHMENTS"

Monks wore their hair in a tonsure where the top of their heads was shaved. Obviously they would need a barber for this and so there was someone associated with the monastery trained in handling sharp blades. Should a patient need to be bled this task was given to the barber and most surgical and dental tasks were also appointed to these barber-surgeons.

The relative wealth of monastic houses allowed them to purchase medical items and build facilities most people could not otherwise afford. Archaeological digs on British monasteries have revealed vessels for distilling medicines, drugs imported from the Mediterranean, and medical tools. In short a monastery could offer a patient both prayers and practical care.

Schools and Training

Monastic settings may have preserved much ancient knowledge but during the medieval period more secular centres of training also appeared. The schools of Salerno and Montpellier taught doctors the ancient classics but also produced new medical treatises. Key to their success was access to Greek, Arabic, and Jewish learning which began to flow into Europe.

Established universities across the continent began to create their own medical schools where doctors could be trained. With training came codes about who could call themselves a doctor. In 1231 Frederick II, Holy Roman Emperor, decreed that all doctors must study logic for three years before undertaking five years of medical study. The doctor then had to pass a test to be granted a license to practice.

Doctors were expected to know the *Ars Medicinae* - 'The Art of Medicine' - which was a textbook composed of ancient works on medical theory, diagnostics, and how to examine everything from a patient's urine to their pulse. As more Arabic texts were translated into Latin they supplemented European medical knowledge. Ibn Sina, an Arabic scholar known in the West as Avicenna, created a canon of medicine that was translated in the 12th century and became a standard text for European doctors.

SERFS AND ROYALS

No one is immune to the dangers of ill health but social class certainly changes the sort of illnesses a person suffers and recovers from. The medieval world was one of strict hierarchy and feudalism which placed people in groups of greatly differing wealth and lifestyles.

Monarchs and nobility were spared the vigorous labour required to earn a living. Archaeological studies of teeth and skeletons, as well as written records, reveal that the wealthy ate a varied diet which contributed to health. Some may have overindulged in meats and cheeses but they were spared the dangers of famines. Wealthier people had greater access to hygiene, baths, and doctors.

For those at the bottom of the social order life was often hungry and short. Their diet consisted mostly of grains and ale for the bulk of their calories. The coarse bread produced at the time tended to grind down their teeth. Signs of poor nutrition are often clearly seen in their skeletons and hard work on farms is also seen from the wear on joints. In times of lean harvests it was the poor who had to go without food. Cemeteries bear witness to the different health outcomes for the rich and the poor.

The disparity in wealth and health between the rich and the poor can be detected in the ailments which their bodies suffered

The works of the ancient Greek Hippocrates were widely copied and sought out by those hoping to undertake medical training

Archaeological evidence shows that many broken bones were set perfectly allowing for complete healing

Medieval doctors were encouraged to treat their patients in a holistic fashion. Based on Galen's teaching they considered what were known as the six 'non naturals': air, diet, sleep, exercise, excretion, and emotion. These were factors that could create either health or illness and so needed to be examined before a treatment could be created for the patient.

Doctors with medical training were separate from barber-surgeons. Any illnesses that required an operation were passed from doctors to their colleagues who used blades. By the 14th century doctors were, on a limited basis, allowed to conduct occasional dissections and so began to have detailed anatomical training they could use for surgery. Some medical schools also demanded that students spend at least some of their time studying surgery.

Access to university-trained doctors however remained far out of reach for the majority of the population. The cost of calling in a doctor and buying the drugs they prescribed was simply too expensive. Fortunately there were alternative sources of medical help.

Superstition and Folk Cures

For those without access to a medically trained healer, most would have relied on traditional cures involving local plants. Bishop wort was thought to have over 50 uses from curing snake bites to preventing drunkenness. Garlic was used for constipation and dropsy (oedema). Some simple remedies would have been preserved within families but when special knowledge was known people sought out wise men and wise women - local healers with a working knowledge of herbalism and charms.

Charms and amulets were a familiar part of medieval life. Some charms were simple, such as hanging a stone with a hole in it to ward off witchcraft, while others were complex, such as written spells worn on the body. The idea of objects offering safety was natural in a world where saints' relics and images were thought to have mystical healing properties.

Some members of a community could genuinely offer sound medical advice in times of need. Midwives were called on to help women during childbirth and their experience no doubt aided in many a dangerous labour. While in towns there may have been professional midwives, for women living in villages childbirth was guided by the older women of the family who had gone through its rigours themselves.

Medieval people should not be looked down on for seeking medical care in any of the forms it existed in at the time. Sickness causes desperation and they were doing the best they could with all the knowledge available to them at the time. Some of their remedies are effective, and still in use now.

NATURAL PHILOSOPHY

BEFORE THERE WAS SCIENCE, THERE WAS NATURAL PHILOSOPHY

WRITTEN BY **EDOARDO ALBERT**

Plato. You've probably heard of him. The greatest of the Greek philosophers. Arguably, the greatest philosopher ever. He formulated the basic questions that philosophers still seek to answer. But while Plato kickstarted metaphysics, epistemology, ethics and political philosophy, he wasn't much interested in the natural world. For him, the world we live in was but a poor, distorted reflection of the world of Forms, where resided the ideal forms of all the things we encounter in daily life: the ideal chair, the ideal loaf of bread, the ideal circle. The things we see around us in everyday life are but imperfect approaches to these ideals.

As such, Plato wasn't much interested in the problems of the natural world, given that the natural world was the realm of imperfect forms. However, Plato's student, Aristotle (who took a side gig as Alexander the Great's tutor) was interested in the natural world. Aristotle was the first natural philosopher. He sought to understand the natural world, defined as the world outside the human sphere, in terms of what he called the four causes. These were the material cause, the formal cause, the efficient cause and the final cause. By interrogating something in relation to these causes, Aristotle sought to establish what something was made from, its structure, what had brought it about, and what its purpose was - or the what, how and why of something (with 'what' taking up the heavy lifting of both material and formal causes).

Aristotle then applied this analysis to things in the world, particularly living things. Unlike Plato, he was an excellent observer of living creatures, taking an intense interest in them. Indeed, his observations of sea creatures would remain the most accurate for over 2000 years.

So while Plato founded philosophy it was Aristotle who was responsible for natural

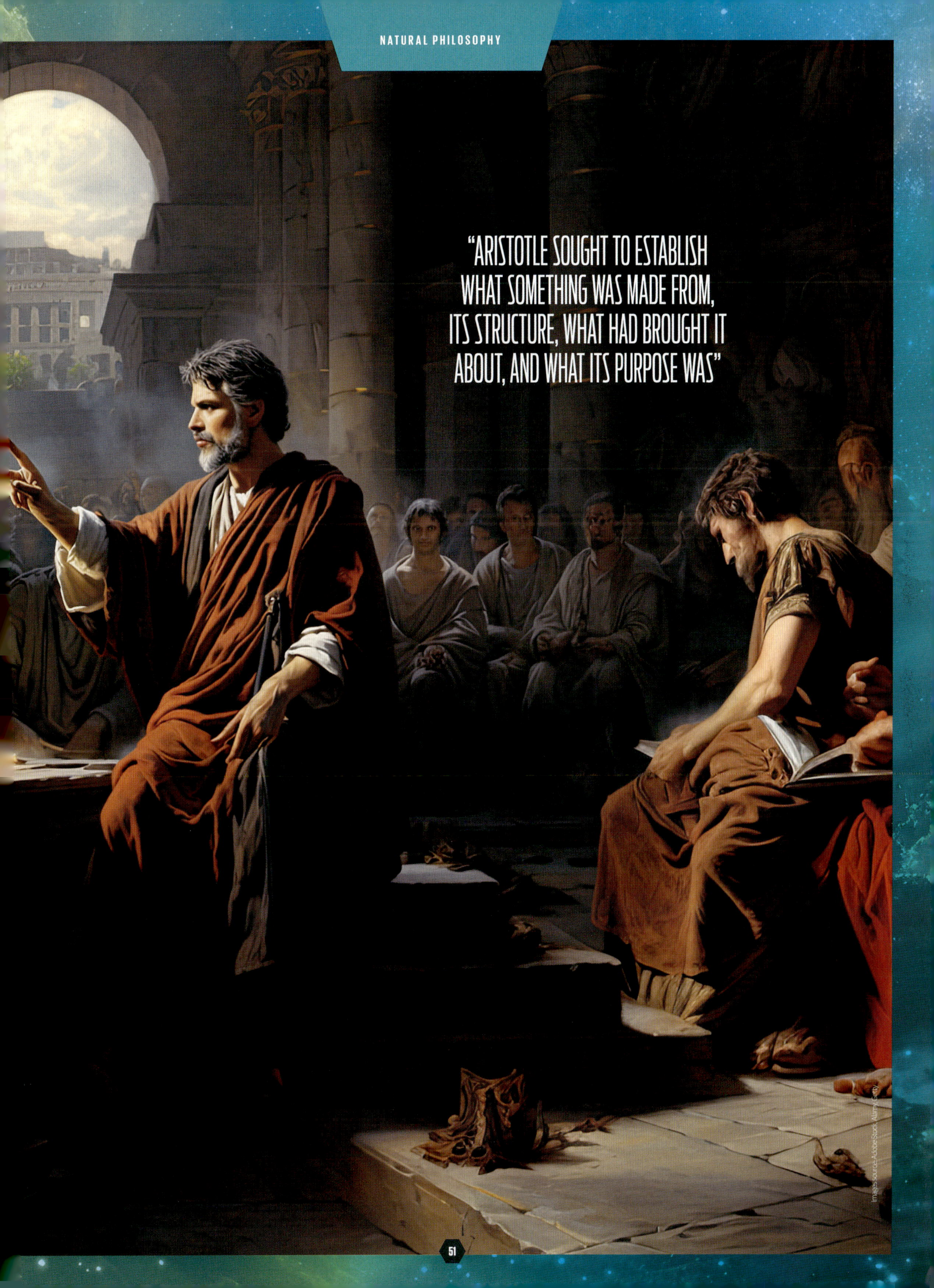

"ARISTOTLE SOUGHT TO ESTABLISH WHAT SOMETHING WAS MADE FROM, ITS STRUCTURE, WHAT HAD BROUGHT IT ABOUT, AND WHAT ITS PURPOSE WAS"

Images source: Adobe Stock, Alamy, Getty

Raphael's famous painting of Plato and Aristotle talking at the school Plato had established in Athens

The greatest of the modern natural philosophers: Sir Isaac Newton (4 January 1643 – 31 March 1727)

philosophy. And it was the sheer sense of wonder at the natural world around him that drove Aristotle to observe, investigate and attempt to understand that world.

To pursue his investigations, Aristotle left Athens and sailed to the island of Lesbos in the Aegean Sea. There, he set about watching the fish and other life that lived near the Gulf of Kalloni, recording his observations in his book, *Historia Animalium*. His observations were detailed and over a long enough period for him to observe migratory behaviour, which he sought to explain through features of the animals' natural lives, namely changes in the amount of food, changing temperatures and the desire to find a mate.

This was as pure an example of early natural philosophy as it is possible to find, for Aristotle sought to explain the behaviour he observed through factors intrinsic to the animals and their environment, rather than by trying to impose upon them some external motivation.

What also stands out in Aristotle is his deep respect for what he actually saw. He was concerned with the truth of his observations and was committed to making them as accurately as possible. What was more, he sought to make his explanations adhere to his observations.

To best understand how Aristotle differed from Plato, we can look at Plato's famous metaphor of the cave. In the *Republic*, Plato writes a dialogue between Socrates and Glaucon where Socrates describes a cave. In that cave there are prisoners who have been kept there from birth, their bodies bound and their heads held immobile so that they can only look at the far wall of the cave. Behind the prisoners, there is a large, bright fire. The prisoners' captors pass between the fire and the backs of the prisoners, and the fire light casts their shadows upon the far wall.

These shadows are the only moving things that the prisoners can see and, not unnaturally, they take them as real. Because the shadows are all they have ever seen, they take them as reality.

"ARISTOTLE SOUGHT TO EXPLAIN THE BEHAVIOUR HE OBSERVED THROUGH FACTORS INTRINSIC TO THE ANIMALS AND THEIR ENVIRONMENT"

But then, one of the prisoners is freed. He looks round but the fire hurts his eyes, eyes that have only ever looked upon shadows. In his pain, he turns back to the shadows, content to look on them.

However, the prisoner is not allowed to rest. One of the captors comes and carries him physically out of the cave, depositing him on the ground outside the entrance. The sun blinds him completely. The wind, the sounds of birds, these all leave him completely disorientated. But he is helpless, he can't find his way back into the cave. And, after a while, his eyes begin to adjust. His vision slowly clears and he looks down and he sees the grass. He looks around and he sees the trees. And then, finally, he looks up and sees the sky and the sun.

The prisoner realises that everything he had seen and accepted as real before were merely shadows of what was truly real. In his excitement, he runs back into the cave to tell the other prisoners what he has discovered. But they don't believe him. They decide that his trip to strange places has driven him mad and they drown out his pleas and stopper their ears. The cave is all the reality they are prepared to see.

The metaphor of the cave is perhaps the richest and most discussed metaphor in the whole history of philosophy. It perfectly sums up Plato's belief that we live in a world of shadows and that what we see around us are but imperfect representations of the true reality of things, which resides in the world of Forms.

Plato taught this theory to Aristotle - and Aristotle totally rejected it. The pupil looked around him, at the world of animals, fish and birds that he observed so minutely, at the trees and the plants, and decided that these were not imperfect representations of some unattainable ideals but real and existing things in their own right. For Aristotle, the

When his son, Alexander, turned 13, King Philip of Macedon started looking for suitable tutors. He chose Aristotle. Aristotle taught the young Alexander and his friends, with them all living together in an early boarding school

particular was the possible, whereas for Plato each particular thing was merely a passport to the universal.

Interestingly, these conflicting ideas went on to play a key role in the intellectual tension that lay at the heart of natural philosophy and which helped give rise to science. For Plato's search for the universal came to underlie the quest for universal laws by which we can understand the universe around us, while Aristotle's emphasis on the particular led to theory being produced from, and tested by, empirical evidence.

This openness to the natural world received a huge boost some 700 years later. The single most important thinker of the early Christian Church was an African, a man who had been a noted libertine in his youth and who had dabbled in the fashionable religions of the late Roman Empire before becoming a Christian. His name was Augustine and he became bishop of Hippo in what is today Algeria.

According to Augustine, God gave human beings two books by which to understand him. One was scripture, the Bible. But the other was the metaphorical book of nature - the world around us. As the two books are both written by the same author, their contents cannot contradict each other. If they do appear to contradict each other, said Augustine, then we need to look deeper and better at scripture and nature in order to unlock the apparent contradiction.

Natural philosophy became the main way for looking at the natural world, bringing to bear upon the question tools from disciplines as diverse as theology, logic and aesthetics, as well as observational skills. For a thousand years, theology was the queen of the sciences, where science meant knowledge, and natural philosophy was her main servant. The two worked harmoniously together.

In fact, their partnership reached its zenith in the work of Isaac Newton. His great work, *Philosophiae Naturalis Principia Mathematica*, translates as the 'Mathematical Principles of Natural Philosophy'. Newton took the precise observations of the movement of objects and universalised them, showing that the law that caused an apple to fall from a tree also moved the Moon and the Earth. Newton brought about the union of the approaches of Plato and Aristotle, extracting the universal from the particular. And Newton was very much a natural philosopher, combining what we would call his scientific work with theology and alchemy. The irony is that such was Newton's success, and so great his prestige, that his successors ignored the philosophy that underpinned his work and focused only on the science. Natural philosophy became a short sub-heading in books about the history of science, while science itself split into more and more subjects, disappearing further into the thickets of specialisation.

It was not what Newton, or Plato, or Aristotle, would have wanted.

THE REBIRTH OF NATURAL PHILOSOPHY

By the 20th century, science had completely split from natural philosophy. The two subjects, so long tied so fruitfully together, became increasingly specialised. Science split into different subjects, which spawned sub-divisions, each so specialised that it became increasingly difficult for the different subjects to speak to each other. Philosophy disappeared down a navel-gazing hole that led, finally, to the very words it used for its discourse dissolving. Now however, influential scientists and philosophers have started arguing that the divorcees would benefit from getting back together. Natural philosophy, in this view, allows scientists to centre their science in a wider world of ideas that partake of the quest for humanity's place in the universe. Not only would the rebirth of natural philosophy correct science's divorce from questions of ethics and virtue, but it would also enable philosophers to engage in what is really their core pursuit: the rational inquiry into the fundamental questions of life. The divorce between science and philosophy stopped scholars in both disciplines from thinking about these big questions. Their reunion could lead to a rebirth for both.

Natural philosophy is ultimately grounded in a profound sense of wonder at this extraordinary world in which we find ourselves

Plato believed that the natural world was less worthy of study

NICOLAUS COPERNICUS

HOW THE OBSERVATIONS OF A 16TH-CENTURY POLISH CATHOLIC MONK SET THE SCIENTIFIC REVOLUTION INTO MOTION AND REALIGNED EARTH'S PLACE IN THE UNIVERSE...

WRITTEN BY **BEN GAZUR**

Until Copernicus, the most widely accepted theory of the universe was that the Earth was at the centre, and the Sun, Moon and planets all revolved around it. This was an idea that had been advocated by Aristotle millennia before, then by Ptolemy, and was fiercely backed by religious leaders in the Catholic Church. Any who dared to challenge this dogma were accused of heresy - a crime punishable by death.

However, following the Ottoman conquest of Constantinople - a bastion of Greek culture - many of the city's scholars were forced to flee west. With them they brought a wealth of ancient knowledge and classical methods of observation and questioning as a way to solve the great mysteries of the universe.

At the same time, growing scepticism of the Catholic Church and England's break from Rome meant that reason was beginning to take the place of religion in academia, a realm that had before been largely governed by Christian belief. As a result, Renaissance astronomers began to challenge Aristotelian physics.

One of these stargazers was Nikołaj Kopernik, or as he has come to be known, Nicolaus Copernicus. He was born in 1473 in Torun, Poland. When he was just ten years old, his father died and his uncle, a bishop, took the boy under his wing. He supported Copernicus throughout his studies at the University of Krakow, which he began in 1491, in the heyday of the Krakow astronomical-mathematical school. It was here that Copernicus laid down a strong foundation for his later mathematical achievements.

In 1496, he moved to Bologna, Italy, to study canon law, and rented a room in the house of prominent professor and astronomer Domenico Maria de Novara. He became his disciple and assistant and for the first time was met with a mind that dared to challenge the existing theories of the cosmos.

A LIFE'S WORK

WE PICK OUT SOME OF THE MILESTONES OVER THIS ASTRONOMER'S LIFETIME

1473
Copernicus is born to his namesake father Nicolaus, an affluent copper merchant, and mother Barbara in Torun.

1491
Copernicus studies painting and mathematics at the University of Krakow at 18 years of age.

1496
After graduating he moves to Italy to study canon law, but is drawn towards astronomy.

1503
Copernicus returns to Poland to live with his uncle, acting as his secretary and physician.

1512
After his uncle's death, Copernicus moves to Frombork (pictured right) in northern Poland to work as a church canon.

TOP 5 FACTS: NICOLAUS COPERNICUS

1 Multitalented man

Copernicus's skills were not limited to astronomy. He was highly educated and was also a physician, scholar, economist, translator, mathematician, artist and diplomat, among other things.

2 It's in the chemistry

The chemical element Copernicium is named after Copernicus. Its discoverers wanted to name the element after a scientist who did not receive enough recognition for their work in their own lifetime.

3 Bad money drives out good money

In 1526, Copernicus developed a monetary theory, now called Gresham's Law, which was used to stabilise the currency in Poland and is still a principal concept in economics today.

4 The centre of everything?

Contrary to popular belief, Copernicus didn't actually believe that the planets of the Solar System orbited the Sun itself, but around a centre that was near to it.

5 Doctor Copernicus

Though he had a brief stint studying medicine, he never gained a medical degree, yet he acted as physician to his uncle and then his uncle's successor for many years.

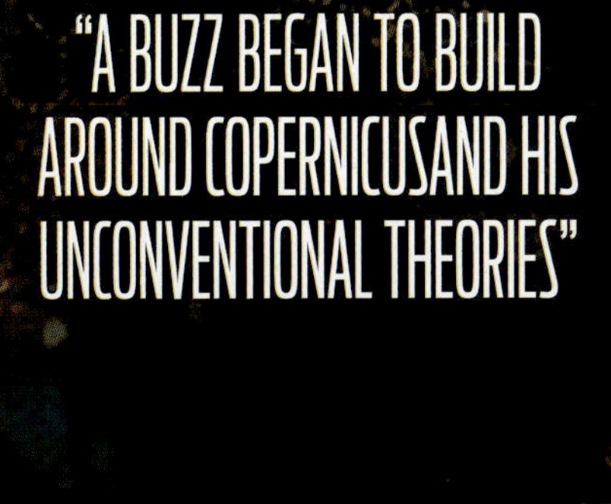

"A BUZZ BEGAN TO BUILD AROUND COPERNICUSAND HIS UNCONVENTIONAL THEORIES"

On completing his studies, Copernicus returned to Poland to live with his uncle, acting as his secretary and physician at the same time. During this time he began work on his now famed heliocentric theory. In 1512, his uncle died and Copernicus moved to Frombork, where he took up a position as a canon - an administrative appointment in the Church.

Here he had more time to devote to his astronomical studies, and built himself a small observatory from which he could plot the movement of the stars. Around 1514, he outlined his theories in a short, anonymous manuscript referred to only as *Commentariolus*, in which he summarised his heliocentric model of the Solar System, where the planets orbited the Sun. While he only distributed the manuscript among a few friends, a buzz began to build around Copernicus and his unconventional theories, but his new ideas also sparked controversy in the Catholic Church.

A 17th century engraving of Copernican heliocentrism

While the threat of persecution did not deter Copernicus from developing his theories, he was reluctant to publish them and kept his findings secret for decades. However, in 1540 his pupil Georg Joachim Rheticus convinced him to publish his book *De Revolutionibus Orbium Coelestium* ('On The Revolutions Of The Heavenly Spheres'). In 1543, as Copernicus lay on his deathbed, the first-ever printed copy was placed into his hands. The Scientific Revolution had begun in earnest.

1514
Copernicus produces a short manuscript known as the *Commentariolus*, where he outlines his heliocentric theories.

1532
Copernicus completes *De Revolutionibus Orbium Coelestium*, but is reluctant to publish it for fear of religious persecution.

1539
The astronomer takes on his first and only pupil, Rheticus (right), who persuades him to publish his work.

1543
The first copy of his book is published, but Copernicus dies shortly after from a stroke, aged 70.

Images source: Wiki, Adobe Stock

THE THIRST FOR SCIENTIA

DURING THE RENAISSANCE, MARINERS BEGAN TO DISCOVER THE SIZE OF THEIR PLANET AND SCHOLARS BEGAN TO SPECULATE ABOUT ITS LOCATION IN SPACE

WRITTEN BY **DEREK WILSON**

In the closing years of the 15th century, the question that was exciting and intriguing thinking Europeans was, 'What is this 'Earth' on which we are standing?'. Within a single decade, 1492-1502, knowledge of the planet had been revolutionised. Before that all geography was based on the *Geographia* written by Claudius Ptolemy in the 2nd century CE. It mapped the Roman world as it was known at the time from the eastern fringe of China to the Canary Islands. Ptolemy knew (contrary to common legend) that the Earth was a sphere but the 'missing' three quarters was inaccessible without ships able to cross the Atlantic or circumnavigate the landmass of Africa. In this remarkable decade, Christopher Columbus explored the West Indies and Central America; Pedro Cabral landed in Brazil; John and Sebastian Cabot found the coast of North America; and Vasco da Gama rounded the Cape of Good Hope and went on to establish a trading colony in India. By 1522, the first Europeans had sailed right around the world.

This opened up hitherto undreamed of opportunities for trade, colonization and exploitation of the resources of distant lands. It also presented challenges to astronomers, mathematicians and inventors. There was a sudden demand for charts, maps and navigational instruments. One of those who shared the new interest in the study of the movements of heavenly bodies was Nicolaus Copernicus (1473-1543), a minor Church official. This polymath, who had collected qualifications in major disciplines from various universities, had a passion for mathematics and astronomy. He applied his formidable mind to solving one of the biggest problems that had been facing scholars over many centuries.

Conventional thinking about the universe was based on Ptolemy's *Almagest*, which placed the Earth at the centre, with the Moon, Sun, nearer planets and distant heavenly bodies revolving around it in concentric circles. By means of painstaking observations and calculations, Ptolemy had provided tables for calculating the position of the planets at any given time. The problem was that not all the gathered data agreed with planetary motions that were actually observed. Copernicus now gave his attention to another ancient theory, that the Sun was the centre of the universe and everything else, including the Earth, revolved round it. He acknowledged that this 'heliocentric' theory did not solve everything but suggested that it was more plausible than Ptolemy's 'geocentric' theory. The book detailing his findings, *De Revolutionibus Orbium Coelestium*, was published shortly after his death.

Opponents offered several objections. Some said that the idea of Earth spinning through space was against common sense: if it were true, violent winds would be constantly ripping across the surface, flinging off all movable objects. Some cited the Bible, which taught a geocentric divine plan. Indeed, the Book of Joshua even spoke of the Sun being made to stand still for a whole day. Any kind of proof of this wild idea rested on mathematical calculation and not observation. That is why final resolution of the problem had to wait until the invention of the telescope in the 17th century.

JOHN DEE 1527 - C.1608

The interests of the remarkable polymath, Dr John Dee, touched several aspects of Renaissance thought and practice - astrology, geography, alchemy and politics. Mathematics held an almost magical fascination for him; Pico della Mirandola observed, "By number, a way is had to the searching out and understanding of everything able to be known." Believing that all scholarly research should be useful, Dee spent time in Louvain with Gerard Mercator, the father of modern cartography who produced several new maps which laid the Ptolemaic system to rest. At a time when England was entering the colonial race in competition with Spain and Portugal, Dee was welcomed at the court of Elizabeth I. He produced charts, navigational instruments and astronomical tables and, in his General And Rare Memorial Pertaining To The Perfect Art Of Navigation (1577) encouraged English overseas expansion. But astrology and alchemy began to take up more of his time. Together with his assistant, Edward Kelley, he attempted to conjure spirits and he spent several years touring European courts as a magician. But he fell foul of popular suspicion. While he was out of the country his neighbours trashed his laboratory and looted his library. In the course of time Dee's influential patrons deserted him and, by the time of his death, he was virtually destitute.

For many years Dee was the queen's astrologer and frequently cast her horoscope

Images source: Wiki

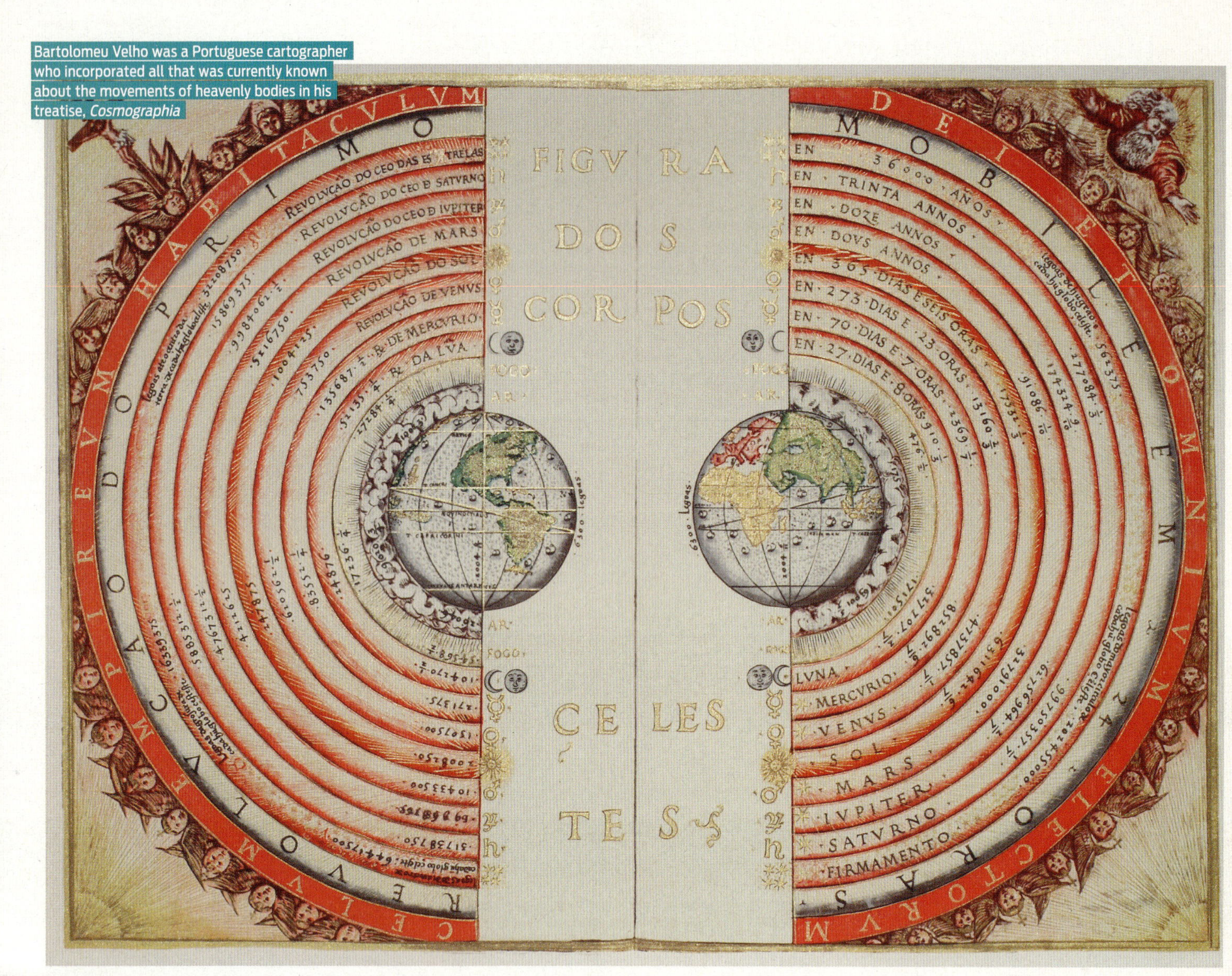

Bartolomeu Velho was a Portuguese cartographer who incorporated all that was currently known about the movements of heavenly bodies in his treatise, *Cosmographia*

In this late Renaissance engraving, the artist points out the folly of get-rich-quick alchemists, seeking the philosopher's stone which will turn base metal into gold

Getting the movements of heavenly bodies 'right' mattered, not only for such activities as maritime navigation but also for the daily life of all people. There was no distinction between astronomy and astrology. The spiritual, intellectual and physical aspects of human experience were bound up together. The ultimate sphere of Ptolemy's system was heaven, from where God's influence permeated down through all the other levels. One of the more controversial Renaissance thinkers, Henry Cornelius Agrippa (1486-1535) expressed it thus in his *De Occulta Philosophia in 1533:*

"The very original and chief worker of all doth by Angels, the Heavens, Stars, Elements, Animals, Plants, Metals and Stones convey from himself the virtues of his Omnipotency upon us, for whose service he made and created all these things."

No-one doubted that the movements of the heavenly bodies of the Solar System (within

Paracelsus was a vital figure in getting thinkers to move away from alchemy and towards chemistry

Uraniborg was dedicated to Urania, the Muse of Astronomy

the sphere known as the zodiac) influenced human destiny. Therefore, the casting of horoscopes was important. It formed a vital part of medical practice and everyone who could afford it, including rulers and Church leaders, employed astrologers, so that all of their decisions and activities could be timed to coincide with the most 'auspicious' conjunctions of heavenly bodies.

The Danish scholar, Tycho Brahe (1546-1601), was still grappling with the same problem 50 years later. This wealthy aristocrat and amateur astronomer was an extrovert showman. He threw lavish parties, kept a beer-swilling elk indoors and wore a metal sheath over his nose to cover a fencing scar. But he had a serious side, studied at various universities, and devoted years of exhaustive labour to probing the heavens and measuring the movements of the major heavenly bodies. He decided the only way to reach conclusions was by plotting the orbits of planets and their relations to one another. Thanks to the support of Frederick II of Denmark, Tycho had the resources for this laborious work. He was given the island of Hven, between Denmark and Sweden, and generous funds to set up a complex called Uraniborg, consisting of a library, laboratory, workshops and a printworks to enable him to published his findings. Here he employed an army of virtually slave workers who created, to his specifications, a large number of quadrants, sextants and other instruments.

Tycho was a firm believer in astrology and his principle motive was to plot planetary movements as precisely as possible so that horoscopes could be drawn with maximum accuracy. But for all his efforts he could not explain the inaccuracies which beset both the Ptolomeic and Copernican theories. Eventually, he produced his own compromise solution - a geo-heliocentric theory. He suggested that, while Sun, Moon and farther stars circled the Earth, the five planets of the Solar System revolved around the Sun. Unfortunately for Tycho, in 1597 his patron died and the new king did not like him. The master of Uraniborg was forced to leave. As soon as he did so, the buildings of the hated task-master were destroyed by his resentful employees. And still the world was waiting for the telescope.

Another branch of ancient wisdom given a new twist by men of the Renaissance was alchemy. The basic belief of alchemists was that everything in the universe was made up of four elements - earth, air, fire and water. This theory had implications for human wellbeing as well as for physics. In medical practice the four elements were linked to four bodily 'humours' governed by phlegm, blood, yellow bile and black bile. Good health consisted in keeping the humours in balance. The application of alchemy to physics followed much the same way of thinking. As all substances were made from the same four ingredients, just in different proportions, it was possible to change their make up (by heating, distilling, mixing, etc). Thus, for example, base metals could be changed (transmuted) into gold. Unfortunately these basic theories did little for the progress of medical science and lent themselves to abuse by fraudsters.

One Renaissance thinker who challenged the use (though not the importance) of alchemy was Philippus Aureolus Bombastus von Hohenheim (1493-1541), who adopted the name of Paracelsus. This aggressively controversial German lived the life of a wandering scholar and writer and was a formative figure in the transition from alchemy to chemistry. He asserted that the human body is a bundle of chemicals. Since it is also bound up with the rest of creation, it was clearly logical to seek remedies for ailments in the animal, vegetable and mineral kingdoms.

It was also Paracelsus's belief that nature abounded in 'similars'. For example, afflictions of the ear could be treated with distillations of cyclamen, whose leaves are similar to the ear. His wide-ranging experiments and observations led him along several false trails but he did make some useful discoveries. He insisted that mind and body were both essential in treating ailments, thus linking the ancient conviction that man is an immortal soul in a mortal body with the later concept of holistic medicine.

DEVILISH DEALS

The Tragicall Hiſtory
of the Life and Death

LONDON,

The thirst for knowledge (scientia) which drove the Renaissance posed ethical and spiritual problems. The exploration of other religions and philosophies were seen by some Church leaders as bringing Christian truth into question. More dangerous, however, was the study and practice of magic. Some scholars (or magi) went beyond astrology and attempted to make contact with beings of the heavenly realm. It was, then, a short step to harnessing the power of such contacts to control terrestrial events. John Dee devoted much of his life to attempting to conjure spirits. Most Renaissance thinkers warned against such activity, pointing out that legitimate scientia had its limits. Early in the 16th century a cautionary tale began to spread about 'Dr Faustus' (or Faust). Whether it had a factual basis we do not know but it certainly appealed to those who were suspicious of the claims made by magi. This Faustus had made a pact with the devil, handing over his immortal soul in exchange for worldly wealth and power. In 1587 an anonymous book, *Historia von D Johann Fausten*, was published in Frankfurt. On this the English playwright, Christopher Marlowe, based his drama, *The Tragical History Of The Life And Death Of Dr Faustus*. The play opens with the brilliant scholar dismissing the study of philosophy, medicine, law and theology in order to devote himself to necromancy. "O, what a world of profit and delight, of power, of honour, of omnipotence, is promised to the studious artisan," he exclaims, before conjuring up the demon, Mephistopheles, to do his bidding. The price he pays is eternal damnation.

Brahe was the last of the major naked eye astronomers, working without telescopes

TOP 5 FACTS: TYCHO BRAHE

1 Naked eye
Brahe was the last of the major naked-eye astronomers, as it wasn't until seven years after his death that the first telescopes came into use.

2 Hard nosed
At the age of 19, Brahe lost the bridge of his nose in a sword fight with a fellow student. For the rest of his life he wore a metal prosthesis over it.

3 Tycho the tyrant?
It is rumoured that Brahe led an oppressive regime on the island of Hven, and that he was deeply despised by the people living there working for him.

4 Murder mystery
It was suggested that Brahe had been poisoned, but after being exhumed from his grave in 2010, results indicated that he probably died from a burst bladder or similar ailment.

5 Lunar legacy
Brahe lives on among the stars – literally. The crater Tycho on the Moon is named after him, as is the crater Tycho Brahe found on Mars.

A LIFE'S WORK

A QUICK GUIDE TO TYCHO BRAHE'S ILLUSTRIOUS CAREER AS AN ASTRONOMER

1546
Tycho Brahe is born at Knutstorp Castle in the then-Danish Scania, to nobleman Otte Brahe and his wife Beate Bille.

1559
Brahe begins his studies in law at the University of Copenhagen.

1560
The prediction of a solar eclipse on 21 August 1560 impresses Brahe enormously, and inspires him to study astronomy.

1572
Brahe first observes a new star, now known as SN 1572, from the Herrevad Abbey observatory.

1573
Brahe publishes his book, *De Nova Stella*, coining the term 'nova' for a new star.

TYCHO BRAHE

MEET THE MAN WHO COINED THE TERM 'NOVA' AND CALCULATED PLANETARY MOTION BEFORE TELESCOPES

Few other naked-eye astronomers have plotted the movement of planets quite as accurately as Danish nobleman Tycho Brahe. His observations of a new star in 1572 and the Great Comet of 1577 helped to shake off the Aristotelian belief that the planets and stars were unchanging and locked in 'immutable' celestial spheres.

Brahe's schooling began at an early age. At just two years old, he was taken from the family home by his uncle to start his education. At age 12, he began studying law at the University of Copenhagen, as was the norm for sons of nobility. However, while the solar eclipse of 1560 cast a dark shadow across the Earth, it lit Brahe's passion for astronomy, and he immersed himself in the works of the great astronomers of the time, eagerly devouring all the research he could.

For some time Brahe studied abroad, but upon his return another uncle - Steen Bille - funded the construction of an observatory and chemical laboratory at Herrevad Abbey. It was here in 1572 that he first noticed the appearance of a very bright star. At the time, the popular theory was that the planets and stars were carried on material spheres that fitted tightly around each other. Brahe's observations proved that his sighting was indeed a new star and not a local phenomenon, and therefore this arrangement was impossible. A year later he published *De Nova Et Nullius Aevi Memoria Prius Visa Stella* ('On The New And Never Previously Seen Star') and it was from this that the term 'nova' came into common use to describe the appearance of a new star.

> "WHILE THE SOLAR ECLIPSE OF 1560 CAST A DARK SHADOW ACROSS THE EARTH, IT LIT BRAHE'S PASSION FOR ASTRONOMY"

After another tour abroad, King Frederick II, desperate to keep Brahe in Denmark, offered him the island of Hven and funding to set up another observatory. In 1576 Uraniborg was built, and later an underground observatory called Stjerneborg. As well as being observatories, they also functioned as workshops where Brahe designed and built new instruments. With these he was able to make incredibly accurate observations and the precision of his celestial positions was said to be more accurate than any before.

When King Frederick died in 1588, Brahe's popularity declined. In 1599, after falling out with the new king, Christian IV, Brahe left Denmark and moved to Prague (which was then part of Bohemia). Sponsored by Bohemian king Rudolph II, he built a new observatory at Benátky nad Jizerou. Here he was responsible for compiling the Rudolphine Tables - astronomical tables that would allow calculations of the planetary positions for any time in the past or future. Here Brahe also met Johannes Kepler, who came to be his assistant until Brahe's death in 1601. He entrusted the continuation of his extensive research to Kepler, who published the finished astronomical tables 26 years later.

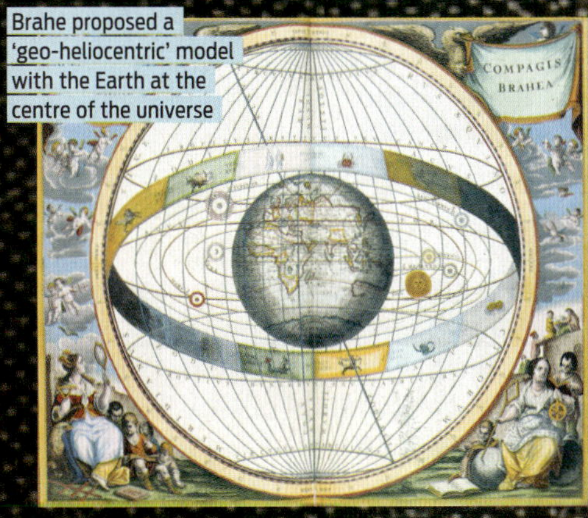

Brahe proposed a 'geo-heliocentric' model with the Earth at the centre of the universe

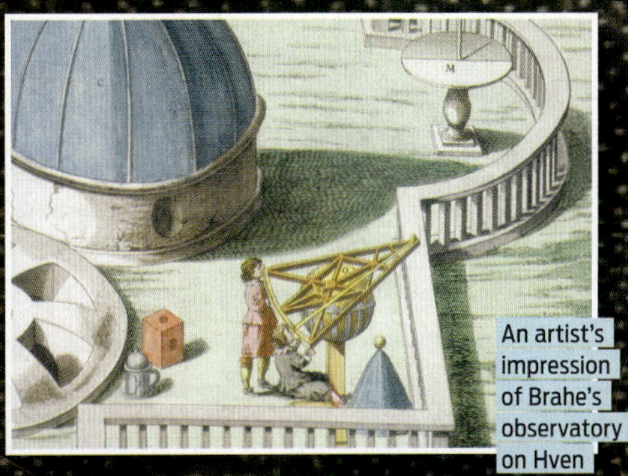
An artist's impression of Brahe's observatory on Hven

1576
King Frederick II of Denmark offers Brahe the island of Hven, where he builds the Uraniborg observatory.

1577
Brahe's observations of the Great Comet (above) prove that objects can move through the celestial spheres.

1599
After a disagreement with the new Danish king Christian IV (right), Brahe moves to Prague, becoming Bohemia's official imperial astronomer.

1601
Brahe suddenly contracts a kidney or bladder ailment and dies 11 days later, aged 54.

Images source: Wiki, Adobe Stock

GALILEO GALILEI

THE FATHER OF MODERN SCIENCE AND ONE OF HISTORY'S MOST INFLUENTIAL FIGURES, TODAY'S ASTRONOMERS OWE GALILEO A GREAT DEBT

Had you been alive in the late-16th and early-17th centuries, Galileo would have challenged, if not changed, the way you looked at the world. His studies into the laws that govern motion, strength of materials and the very nature of scientific method of the time paved the way for scientific advances for the next few centuries. Though the achievement he's best known for was to advocate the heliocentric system, he was such a staunch proponent of this in the face of punitive opposition that the scientific community was forced to re-examine its beliefs.

The world that Galileo was born into in 1564 was as much a boon to his career as a hindrance. On the one hand, contemporary Renaissance-era geniuses like Nicolaus Copernicus and Leonardo da Vinci had already proved the transition between the expanding definitions of the sciences. Italy was a thriving hub for artists, explorers, mathematicians, writers, inventors and more; ideas disseminated with unprecedented freedom and new concepts bubbled up from archaic beliefs, rocking theories of the time that had gone unchallenged for hundreds of years.

On the other hand, Galileo was a tenacious antagonist who lived in Pisa, Italy, at a time when Rome's political power was still very strong and religious censorship was rife. His feud with the Vatican dictated the last few decades of his life, and arguably brought to an end Galileo's world-changing run of stellar discoveries prematurely.

In 1588, at the age of 24, he was already a mathematician of some renown in Italy, having circulated his theories on weight and the centre of gravity while lecturing to the prestigious Florentine Academy. It brought him to the attention of the University of Pisa in 1589, which appointed him the chair of mathematics. It was here that he performed his experiment from the top of the Leaning Tower of Pisa, dropping various weights to the ground and proving that the speed of an object's fall is not proportional to its weight. The backlash against his attack on Aristotle's time-honoured theories saw him released from his position in 1592, although he immediately moved on to greener pastures as chair of mathematics for the University of Padua - part of the Venetian Republic. During his time here he would make several contributions to science that would revolutionise astronomy.

Galileo has been so frequently associated with the telescope that he's commonly

A LIFE'S WORK

KEY POINTS FROM GALILEO'S LIFE AND HIS BRILLIANT, CONTROVERSIAL CAREER

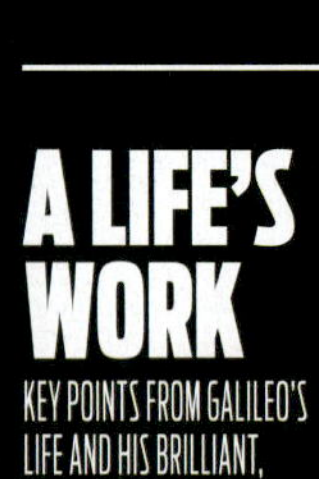

1564 Born 15 February in Pisa, Italy, a city he would return to later in life.

1581 Enrols at the University of Pisa to study medicine, but later decides to study mathematics and philosophy.

1588 Applies for the chair of mathematics at the University of Bologna but doesn't get it.

1592 Galileo's patrons secure him the chair of mathematics at the University of Padua.

1609 Reinvents the telescope and receives substantial financial reward from the Venetian Senate.

TOP 5 FACTS: GALILEO

1 Much of Galileo's work was withdrawn and banned during the 17th century by order of the Church. It wasn't until 1718 that reprinting was allowed again.

2 In addition to the objective lens and thermoscope, Galileo also invented a geometric compass, a microscope, a pendulum clock and contributed greatly to many other technologies too.

3 In 1638 - towards the end of his life - Galileo went blind. Yet even in his final few years he continued with his work, taking on an apprentice to help him who was with Galileo until his death.

4 Sometimes Galileo was far from being correct. For example, he disagreed with Kepler's theory that the Moon caused the Earth's tides and believed that they were down to the rotation of the Earth and orbit of the Sun.

5 Galileo discovered the Jovian moons Io, Callisto, Ganymede and Europa. He named them the Sidera Medicea (Medicean stars) after his patron Cosimo II de' Medici. They would later become known as the Galilean moons.

credited with its invention, which isn't true. The telescope was actually invented in the Netherlands in 1608, proving a watershed for both Galileo's career and science as a whole. He saw how to drastically increase magnification through lens grinding and, in August 1609, he presented his improved telescope design to the Venetian Senate. They were so impressed with his re-invention that they doubled his salary and extended his tenure of the chair of mathematics to a lifetime. This invention was also the tool with which Galileo would achieve his magnum opus.

With a telescope that magnified the sky up to 20 times, he was able to discern celestial objects in unprecedented detail, like the Moon, whose surface he discovered was pocked by craters and not perfectly smooth. He was also able to make out four satellites orbiting Jupiter. This flew in the face of the contemporary Aristotelian thinking at the time: that the Earth was an imperfect and corrupt celestial body surrounded by the immutable heavens. The Moon and the planets in fact revolved around the Sun, which was the centre of the known universe and there was more than one centre of motion within this universe. This revolutionary support of Copernican heliocentrism saw Galileo fall out of favour with the Vatican. After facing an intimidating inquisition in Rome, he was sentenced to lifetime house arrest - a relatively lenient punishment at a time when heresy was usually met with torture, prison or death. Galileo continued his work in secrecy and even managed to smuggle a vital book summarising his research into motion - *Dialogues Concerning Two New Sciences* - out of Italy and have it published in the Netherlands, before he died in 1642.

1610
Makes one of his most famous discoveries - what are now known as the Galilean moons of Jupiter.

1613
Publishes a paper on sunspots, called History And Demonstrations Concerning Sunspots And Their Properties.

1623
Il Saggiatore ('The Assayer') - Galileo's views on physical reality and the scientific revolution - is published.

1632
Publishes his controversial *Dialogue Concerning The Two Chief World Systems*, falling foul of the Church.

1633
After a commission to examine Galileo's work, he is charged with heresy and sentenced to life under house arrest.

1642
In his final years, Galileo summarises his life's work and teaches a student, before he dies.

JOHANNES KEPLER

OFTEN OVERSHADOWED BY GALILEO, KEPLER WAS ONE OF THE MOST IMPORTANT FIGURES IN THE FIELDS OF ASTRONOMY AND PHYSICS

Johannes Kepler was a German mathematician and astronomer who, despite being less well known than scientists such as Galileo during his own lifetime, played a pivotal role in the founding of modern astronomy and how we understand it.

Today, Kepler is best remembered for his three laws of planetary motion, as well as his seminal texts on the orbit of Mars, the shape and formation of planets and the ratification of a Sun-centred model of the Solar System - first posited by Renaissance astronomer Nicolaus Copernicus.

Kepler was born on 27 December 1571 in the free imperial city of Weil der Stadt, near to modern-day Stuttgart. One of his first encounters with astronomy came when he was six, observing the Great Comet of 1577. This was followed three years on with a lunar eclipse, which he later recalled greatly inspired him.

Kepler stayed in touch with astronomy throughout his schooling, retaining his interest during his time at the University of Tübingen. It was at Tübingen where his superb mathematical abilities became evident and he soon gained a reputation as a skilful astronomer and astrologer (in this era, these disciplines were considered the same thing).

Around this time he gained a mentor - Michael Maestlin - and began learning both the traditional Ptolemaic system of planetary motion (which was Earth-centred) and also the Copernican system, which was new and revolutionary and controversially placed the Sun at the heart of our Solar System.

At the age of 23, Kepler started teaching mathematics and astronomy at the University of Graz. It was during his time here that he published the first defence of the Copernican system, his *Mysterium Cosmographicum*. The text was not widely read, but it firmly established Kepler as one of the foremost astronomers of the age, as it largely modernised and honed Copernicus's theories.

In 1600, Kepler met someone who would become a key colleague in the formulation of his three laws: Danish nobleman Tycho Brahe, who was building a new observatory. Here he wished to utilise Brahe's extensive observations of Mars to run a test to back up his evolution of the Sun-centred Copernican system. Following Brahe's untimely death in 1601, he succeeded him in becoming imperial mathematician to Emperor Rudolf II, a time in which he published works on optics and techniques for observing stars

A remnant from a massive supernova observed by Kepler back in 1604

A 31km (19mi)-diameter crater on the Moon that is named after Kepler

A LIFE'S WORK

A JOURNEY THROUGH THE BIG MOMENTS IN KEPLER'S LIFE

1571 Johannes Kepler is born on 27 December in Weil der Stadt, Württemberg, south-west Germany.

1577 Although a sickly child, viewing the Great Comet of 1577 is a turning point that will inspire his career.

1589 At 18 he enrols at the University of Tübingen's stift (the theological seminary).

1594 He starts teaching mathematics at the University of Graz but is dismissed in 1600.

1596 Publishes *Mysterium Cosmographicum*, developing the Copernican theory of heliocentrism.

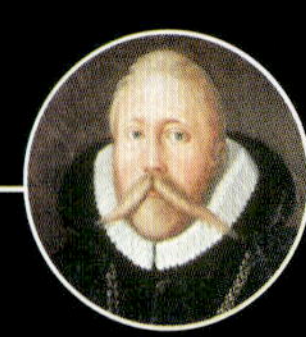

1600 Moves to Prague to work with Danish astronomer Tycho Brahe, who dies the following year.

TOP 5 FACTS: JOHANNES KEPLER

1 Family
Johannes Kepler and his first wife, Barbara Müller, had five children in total. However, the first two – named Heinrich and Susanna – both died in infancy. The following three survived. He married a second time in 1613.

2 Banishment
Kepler's belief in a Sun-centred Solar System – along with his deep-rooted Protestantism – saw him banished from the Austrian, heavily Catholic city of Graz in August 1600.

3 Supernova
Kepler was the first astronomer to observe the SN 1604 star go supernova in October 1604. Two years later he described the event in detail in his text *De Stella Nova*.

4 Rejection
When Kepler published two of his three laws of planetary motion in his groundbreaking work *Astronomia Nova* at first he was ridiculed and ignored by the majority of the scientific establishment, including Galileo Galilei.

5 Mountains
In New Zealand's Fiordland National Park there's a mountain range named after Kepler in tribute to his extensive contributions to the field of astronomy.

"KEPLER IS BEST REMEMBERED FOR HIS THREE LAWS OF PLANETARY MOTION AND TEXTS ON THE RATIFICATION OF A SUN-CENTRED SOLAR SYSTEM"

and planets, as well as his landmark 1609 text *A New Astronomy*, in which he introduced the first two of his three laws of planetary motion.

After the death of the emperor in 1612, Kepler moved to Linz. It was here, seven years later, that he published *Harmonices Mundi*, a text that while filled with much erroneous material as determined by modern science, did include his third and final law of planetary motion. He later completed a comprehensive star catalogue and planetary table that he'd started with Brahe in 1600.

Kepler died on 15 November 1630 in Regensburg, Germany. Despite his impressive work, his three laws of planetary motion were not immediately accepted by the astronomical community, with notable figures such as Galileo and René Descartes ignoring them. It was not until the late-17th century that astronomers like Isaac Newton started to adopt them.

1601
Kepler is appointed Emperor Rudolf II's imperial mathematician in Brahe's stead.

1609
Two of Kepler's three laws of planetary motion are published in *Astronomia Nova*, based largely on the movements of Mars.

1611
In the wake of Galileo's success, Kepler puts forward a new design for a telescope with two convex lenses.

1612
Kepler returns to his hometown to defend his mother, a healer, against charges of witchcraft.

1621
Publishes *Epitome Astronomiae*, which is a compilation of all his other work on heliocentrism.

1624
As per Brahe's dying wish, Kepler completes an astronomical table more accurate than previous models.

1630
Johannes Kepler dies in Regensburg on the way to collect a debt, aged 58.

SIR ISAAC NEWTON

OFTEN CONSIDERED THE FATHER OF MODERN-DAY PHYSICS, MEET ONE OF THE MOST INFLUENTIAL SCIENTISTS OF ALL TIME

> "HE WOULD BE WELCOMED INTO THE ROYAL SOCIETY"

Sir Isaac Newton was an English physicist who laid down the foundations of modern-day classical mechanics. The core of this was his description of universal gravitation and clarifying the existing three laws of motion, which he brought together under one system. This allowed Newton to demonstrate that the motions of celestial bodies were dictated by a single set of universal laws, radically shifting scientific thought away from heliocentrism - the idea of the Sun being at the centre of the entire universe - and setting the stage for Einstein's equally pioneering theories of general and special relativity over 200 years later.

At the time Newton was admitted to Trinity College, Cambridge, in 1661, the university was still basing much of its scientific and mathematical teachings on Aristotle. However, due to Newton's widespread reading of many modern thinkers, the institution was slowly introducing the ideas of Descartes, Kepler and Galileo. He graduated in 1665 and spent the next two years formulating his theories on calculus, optics and gravitation.

Following this, Newton became more interested in optics and he lectured on it between 1670 and 1672. It was during this period that he developed the Newtonian telescope, the world's first functional reflecting specimen, which he presented to the Royal Society alongside an investigation into the refraction of light. He proceeded to conduct much work into the nature and properties of light over the next 30 years, which would culminate in the publication of his 1704 text *Opticks*.

Prior to that, in 1687, Newton published his ground-breaking book *Philosophiae Naturalis Principia Mathematica*, which outlined his laws of motion, universal gravitation and a derivation of Johannes Kepler's laws of planetary motion. Even though his genius had already been noted, this seminal text's success established him in scientific society. He would not only be welcomed into the Royal Society, but also knighted by Queen Anne - only the second scientist to have been awarded the title. Newton also acquired a keen circle of admirers including Edmond Halley.

Newton continued his work in mathematics and science, but also took up the post of warden, then master, of the Royal Mint. In 1703 he became president of the Royal Society and an associate of the French Académie des Sciences.

A LIFE'S WORK

SOME OF THE MAJOR MILESTONES IN NEWTON'S EVENTFUL LIFE

1643
Isaac Newton is born on 4 January in Lincolnshire, England.

1655
Newton attends the King's School from the age of 12 to 17.

1661
He attends Trinity College, Cambridge. After four years, he obtains a degree in maths.

1670
Newton lectures on optics and astronomy at Cambridge University.

1672
Newton builds his famous reflecting telescope (right) and presents it to the Royal Society in London.

TOP 5 FACTS: ISAAC NEWTON

1 Despite Newton's great scientific achievements, he actually wrote more on biblical hermeneutics and occult studies than science. He was a lifelong, if unorthodox, Christian.

2 Newton was only the second scientist in history to be bestowed a knighthood, which he was awarded in 1705. His coat of arms was a shield with two crossed shinbones.

3 Newton was warden of the Royal Mint during the Great Recoinage of 1696. During his time at the Royal Mint he successfully prosecuted 28 forgers for creating illegal currency.

4 In 1704, Newton attempted to glean scientific information from the Bible. From what he extracted from the religious text, he predicted that the end of the world would come no earlier than 2060.

5 After Newton's death in 1727, his hair was found to contain high levels of mercury, indicating he had suffered mercury poisoning.

PHILOSOPHIÆ
NATURALIS
PRINCIPIA
MATHEMATICA

IMPRIMATUR

LONDINI

1687
Newton publishes *Philosophiae Naturalis Principia Mathematica* after years of research into gravitation and planetary motion.

1704
Newton publishes *Opticks*, which demonstrates how a prism can act as a beam expander.

1705
Newton is knighted by Queen Anne due to his scientific work and role as master of the Royal Mint.

1724
With old age and failing health Newton moves in with his niece and her husband at Cranbury Park, near Winchester, England.

1727
Newton dies in his sleep on 31 March. He is aged 84.

Images source: Getty, Wiki, Adobe Stock

ANTONIE VAN LEEUWENHOEK

HE DISCOVERED NEW WORLDS USING HOMEMADE MICROSCOPES

Imagine being the first person ever to see a whole new world that nobody knew existed. In the 1670s, Dutch cloth merchant Antonie van Leeuwenhoek (pronounced 'lay-ven-who-k') did just that. He peered into the realm of mini-creatures, discovering a range of tiny life forms that live everywhere on the planet, including inside our own bodies. At the time, Leeuwenhoek (1632–1723) was the only human to have observed these living things. For this amazing finding, he is called the father of microbiology, kicking off the study of life too small to be seen without microscopes.

Leeuwenhoek didn't set out to be a scientist: he used his microscopes to check the quality of his cloth in close-up detail. In his spare time, Leeuwenhoek painstakingly polished his glass magnifiers to make them better. When he used them to look at pond water, he was amazed to see it was literally crawling with life. Leeuwenhoek called the mini-beasts animalcules, but today we know them to be bacteria (life forms that live in vast numbers almost everywhere on Earth) and protozoa (organisms that are found in most wet places around the world).

He wrote about his discoveries in letters to the Royal Society and was made an honorary member, despite being neither British nor, technically, a man of science. He said of his work, "Whenever I found out anything remarkable, I have thought it my duty to put down my discovery on paper, so that all ingenious people might be informed thereof."

Leeuwenhoek also studied the structure of wood and crystals, discovered red blood cells (which carry oxygen around the body) and made more than 500 microscopes - but he wanted to keep these designs secret. He even spread false rumours about how he did it to put potential competitors on the wrong track. We do know some of his secret methods, though. He used fire to heat glass until it could be pulled into a long, thin thread. When the thread eventually broke, Leeuwenhoek would put the strand back into the fire and mould it into a small glass sphere - making it as small as possible to boost its magnifying effects to new depths.

Without Leeuwenhoek and his homemade microscopes, the discovery of bacteria would have been delayed by at least 100 years.

UNDER THE MICROSCOPE

Leeuwenhoek's incredible handcrafted microscopes pale in comparison with today's modern equipment. Leeuwenhoek could see things as small as 1.35 micrometres (roughly 750 times smaller than a millimetre). With today's technology, we can see objects as small as a nanometre (there are a million nanometres in a millimetre).

Microscopes are essentially tubes with lenses – curved pieces of glass that bend light rays as they pass through. Lenses are everywhere, from torches, LED lights to cameras and in your eye. You can put a single small drop of water on top of the words on this page, to see how a lens bends light to make things appear closer. A typical magnifying glass is a single lens that magnifies by around 10 times. The ones that are used in schools are called compound microscopes and use at least two lenses, which can sometimes magnify objects by 1,000 times or more.

There is a limit to the size of things that can be seen with light, but scientists have created microscopes that replace light with a beam of electrons (a particle with a negative electrical charge), which can see even smaller things.

Nicolaus Copernicus (19 February 1473 – 24 May 1543) was the man most responsible for starting the Scientific Revolution

WHEN THE WORLD CHANGED

HOW THE SCIENTIFIC REVOLUTION CHANGED THE WAY WE SEE AND UNDERSTAND THE WORLD AROUND US

WRITTEN BY **EDOARDO ALBERT**

The strange thing is, it could have been a magical revolution instead. In the 16th century, when the Scientific Revolution kicked off, there was also another revolution going on too, often involving the same people.

In 1492, Columbus had discovered America. A few years later, Vasco Da Gama became the first European to sail round Africa and on to India. In 1522, what remained of Magellan's expedition landed in Spain having circumnavigated the globe. The world was opening up, revealing new and completely unknown lands. This was a tremendous affirmation for the fighting, warring, squabbling European kingdoms, which had for centuries failed to achieve the unity of the Roman Empire. They had now done something that their revered ancestors had never done - never even dreamed of doing. The world was suddenly new, fresh, exciting.

The Hunt for Knowledge

This excitement and sense of discovery spread to the intellectual realm. Scholars hunted markets and monasteries, searching out old, forgotten manuscripts, looking for the lost knowledge of the ancients. Book hunters went on expeditions to out-of-the-way places and exotic cities, bringing back their finds for the gentlemen scholars who waited with eagerness to see what they had found.

And among these treasures were books of magic. Contrary to its popular image, the people of the Middle Ages did not particularly fear witches and witchcraft. The Church regarded witchcraft as a delusion, the play of unhinged minds, and had no truck with witch hunts. But as the Renaissance began, magic crept out of folklore and into the cities. Where before wizards and magic had been the stuff of tales and legend, now they acquired the patina of gossip about the neighbours. And that was because there were serious magicians now living in most European cities, carrying out their rituals and studying their grimoires.

For magic, to the Renaissance mind, was a potential path to power: to power over nature and hence power over their fellow men. And power was the currency underlying the new quest for knowledge.

This was no aberration of stupid minds. In the 16th and on into the first half of the 17th century it was not at all clear which path would lead to greater power: science or magic. The humanist scholars who kickstarted the Renaissance in the second half of the 15th century, men such as Marsilio Ficino and Pico della Mirandola, investigated astrology alongside astronomy, practised alchemy as chemistry, and saw the world as an organism that reflected the motion of the heavens. Such an organically connected whole might be affected by many different means, and one of those means was magic.

As Above, So Below

Part of these magical investigations was the seeking of correspondences between different levels of reality. And nowhere was the correspondence more eagerly sought than in the heavens. Astrology was the queen of the sciences; it influenced all the others

Images source: Adobe Stock, Alamy, Getty

and, as a result, astronomical observations became ever more precise and detailed.

It was the increasing precision of these observations that led the Polish Catholic canon, Nicolaus Copernicus, to become dissatisfied with the old Ptolemaic description of the movement of the Sun, Moon and planets. According to that model, these bodies all circled the fixed Earth in ascending spheres, with the stars set upon an outermost, fixed sphere. It had started out as an elegant theory but it had become increasingly convoluted. In particular, the Ptolemaic model had to cope with the occasional strange motions of the planets: sometimes, they seem to move backwards. Indeed, this waywardness gave them their name: planet comes from planetes, the Greek word for wanderer.

To cope with this retrograde motion, astronomers had to posit that each planet followed a little circular path around its own circular orbit: a little circle moving around a big circle (Ptolemy also believed that the heavenly spheres must be perfect, unlike the Earth, and therefore had to move in the perfect shape, the circle).

As observations became more precise, these epicycles became steadily more convoluted in order to cope with the apparent motion of the planets in the sky. To young Copernicus, this increasingly elaborate model clearly did not answer to the overall requirement for perfection in the heavenly realm that he believed must obtain. The planets moved in circles because the circle was the perfect shape, but to have so many of them made the model clumsy. However, it's worth noting that Copernicus did not propose his system because the Ptolemaic system was making incorrect predictions about the motions of the planets. In fact, it predicted planetary motion very well. No, Copernicus' new, heliocentric model of the Solar System resulted from his conviction that the heavens should operate with perfect symmetry and not with this rather half-baked collection of little circles spinning on big circles. We now know that expecting our idea of 'perfection' from the natural world is a false assumption, but Copernicus was working from a hypothesis that assumed a divine order to the universe.

"AS OBSERVATIONS STOOD AT THE TIME, IT WAS NOT AT ALL CLEAR WHICH THEORY WAS CORRECT"

Copernicus had already worked out the outline for his new heliocentric theory of the universe as a young man but he spent much of the rest of his life collecting data and observations to back it up, while also painstakingly writing a book on the subject. Although Copernicus only published the book right at the end of his life, he had circulated a short paper about his ideas to friends who had copied and passed on the ideas, so the idea of the Sun as the centre around which the Earth and other planets revolved was already in circulation.

The Revolution Begins

De Revolutionibus Orbium Coelestium ('On the Revolutions of the Heavenly Spheres') was published in 1543. Copernicus was given the publishers' proofs as he lay on his death bed, so he saw the fruit of his lifetime's work. Scholars date the start of the Scientific Revolution to the publication of *De Revolutionibus* but the curious matter is that it was a revolutionary textbook that almost nobody read. *De Revolutionibus* was, of course, written in Latin but it was also unremittingly mathematical; only the most eminent astronomers and mathematicians of the day were able to follow it. Looking back from our perspective, it's also strange that the people who did read it were clearly far less interested in the heliocentric part of the theory than they were in how Copernicus calculated planetary positions.

The simple fact is that, as observations stood at the time, it was not at all clear which theory was correct: Copernicus's new heliocentric theory or the time-tested Ptolemaic system. To verify which was correct, better observations were needed.

Step forward Tycho Brahe. This Danish astronomer, who lost his nose in a duel as a young man and wore a prosthetic nose for the rest of his life (a brass one for everyday life and a gold one for special occasions) was the best astronomer before the invention

Alchemy, and other occult arts such as astrology, were important elements of the intellectual explosion of the 16th and 17th centuries

THE CHEMICAL REVOLUTION

Chemistry came late to the scientific party. While astronomy and physics had sorted out their methods and gained overall theoretical frameworks by the 17th century, chemistry was still stuck with old ideas. In part this was down to the sheer difficulty of investigating chemical reactions. The stars and planets could be easily seen through a telescope. The laws of motion could be investigated by dropping a ball. But carrying out chemical reactions and measuring the heat generated while also measuring how the mass changed was beyond the instruments available to chemists in the 17th century. Chemists did have a theory for how substances burned, however. They thought that combustible substances contained a substance, phlogiston, that was released when the substance burned. Plants then absorbed the released phlogiston from the air. However, in the second half of the 18th century, the French chemist Antoine Lavoisier challenged this view. A wealthy man, Lavoisier was able to pay for instruments accurate enough to precisely weigh a substance before it was burned and then the products of that burning. He found that, rather than phlogiston being released, mass was conserved. Fire, Lavoisier discovered, came about when oxygen in the atmosphere combined with the burning substance. Oxygen was his singular discovery, and the quantification of chemistry his signal contribution to the chemical revolution. Unfortunately for Lavoisier, he was caught up in another revolution, the French one, and guillotined at the age of 50.

Lavoisier's meticulous experiments were crucial in starting the Chemical Revolution

of the telescope. His observations were typically within one arcminute of accuracy. He compiled a stellar catalogue that was more accurate than anything that had been published before. And, on 11 November 1572, Tycho saw a new star in the constellation Cassiopeia. This was important because, according to Aristotelian theory, the sphere of the stars did not change. But here was a new star where there should be no new stars. To prove that the new star really was shining from the sphere of the stars and not one of the closer spheres, Tycho demonstrated that it had no parallax (that is, relative motion compared to the other stars). It really was a new star.

While Brahe did not accept the Ptolemaic system, neither did he agree with Copernicus's heliocentric model, instead proposing his own system where the Sun and the Moon orbited the Earth and the other planets orbited the Sun.

Although under house arrest, Galileo continued to receive famous visitors, allowing them to use his telescopes. One of his visitors was the English poet, John Milton

The Revolution Gathers Pace

Ironically, it was Tycho's own research assistant, Johannes Kepler, who set the Copernican system on the road to universal acceptance. Using Tycho's unmatched planetary observations, he was able to see that the orbits of the planets were not the perfect spheres expected of them by Aristotelian theory. Instead, they follow elliptical orbits.

Kepler's three laws of planetary motion were deeply based upon his conviction that the physical universe was an image of the spiritual universe. In common with most other astronomers of the era, Kepler spent as much time working on astrology as he did on astronomy - indeed, for Kepler and his contemporaries, these were one subject, not two separate disciplines.

The laws of planetary motion that Kepler propounded brought astronomy and physics together. Now, astronomers began to search for laws to describe the behaviour of the visible universe.

Looking Up

But to derive those laws, better instruments were required. None would be more revolutionary than the telescope. The first telescope was apparently invented in the Netherlands, although we don't know who the inventor was. However, news of this extraordinary instrument that brought distant things nearer rapidly spread and one of the people who heard about it was a young Italian named Galileo Galilei. Working from a rather garbled description of what went towards a telescope, Galileo set about making his own version, and rapidly succeeded. Having an eye to the main chance, Galileo presented his telescope to the Venetian Senate, who promptly rewarded him with a sinecure at the University of Padua. This telescope magnified objects by eight times but Galileo soon improved his design, making instruments with a magnifying power of twenty and thirty fold.

Then, on 30 November 1609, Galileo pointed his telescope up at the night sky. It was possibly the single most momentous realignment of an instrument in history. Over the course of the next few months Galileo made a series of revolutionary astronomical discoveries.

According to Aristotelian theory, the Moon was perfect. But through his telescope Galileo could clearly see that it had mountains, like the Earth. What's more, he realised that the Moon shone by light reflected from the Earth. Turning his telescope to the other planets, Galileo saw that Venus, the brightest of all the planets from Earth, has phases like the Moon. These phases could only exist if Venus orbited the Sun. And finally, looking at Jupiter, he saw that it had its own satellites too.

Unlike Copernicus, Galileo rushed to publish his findings. Unfortunately, having had the pope, Urban VIII, as a friend and supporter, Galileo seemed to lampoon him as a buffoon in his book, *Dialogue Concerning the Two Chief World Systems*. With the outraged Urban VIII feeling that he had been betrayed, Galileo fell foul of other enemies and his work was condemned and he himself confined under house arrest. It was, however, a fairly mild punishment: Galileo could have visitors and, later, leave the house for visits to his children.

Besides, for his part Galileo saw that his view was gradually being vindicated. It was becoming clear that the Earth did indeed move around the Sun. The old Aristotelian view of the universe was shattering, the

EXPERIMENTER TO THE END

Francis Bacon, the man who set the programme for the Scientific Revolution in his book, *Novum Organum*, experienced the highs and lows of public office. During the reign of James I, Bacon was knighted, made Attorney General and finally raised to the office of Lord Chancellor. This was a prestigious and powerful position, but also one in which Bacon, who was perpetually short of money, was exposed to the temptation of bribery. It was a temptation he did not always resist and, in 1621, Bacon's enemies struck. He was charged with corruption, found guilty, fined £40,000 and sent to the Tower. King James, mindful of Bacon's previous service, commuted the fine and released him from the Tower, but Bacon's public life was done. The public disgrace did not stop his research, however, and Bacon continued to work and write. Then, in March 1626, while driving in a carriage to Highgate (then a village north of London), Bacon saw the opportunity to try another experiment. It had been a harsh winter and snow still lay on the ground. Seeing it, Bacon wondered if the snow and ice would help preserve meat. So he ordered the coach to stop. He got out, bought a chicken from a local woman, had her gut the bird and then Bacon proceeded to stuff the dead animal with snow. However, Bacon became so chilled by the experiment that he fell ill and was taken to the nearby house of the Earl of Arundel. There, he was put to bed, but the chill he took soon developed into bronchitis and he died two days later, an experimenter to the end.

Francis Bacon (22 January 1561 – 9 April 1626) was a true Jacobean gentleman, complete with ruff and beard

perfect crystalline spheres breaking apart and opening up a view of a much larger universe. But it was a universe governed by mathematical, not divine, laws.

Power it Up

While Copernicus, Kepler and Galileo were the footsoldiers of the Scientific Revolution, it was an Englishman who was the general, writing the book that became the manifesto for a new ideology of knowledge. That Englishman's name was Francis Bacon. Bacon was a true Renaissance man, writing on subjects as diverse as the scientific method and the law, while also maintaining a parliamentary career during the reigns of Elizabeth I and James I. Somewhat unwittingly, centuries after his death Bacon also became the go-to choice for people who believed that a rural hick from the Midlands who never went to university could not possibly have written *Hamlet*.

While Francis Bacon didn't write Shakespeare's plays, he did write the manifesto of the Scientific Revolution. The *Novum Organum* was published in 1620, some 12 years before Galileo's *Dialogue*. The *Novum Organum* had its genesis in Bacon's time studying Aristotle, whom he revered

The Copernican system placed the Sun at the centre, with the Earth and the other planets orbiting around it, although Copernicus thought their orbits were perfect circles

greatly but whose philosophy he became increasingly impatient with. In particular, Bacon came to the conclusion that the efforts of natural philosophers to understand the world in terms of Aristotle's Four Causes was fundamentally misguided. He argued that efforts to find the final cause of something, its purpose, had served solely to foster dissent and argument among natural philosophers. He proposed that they should forthwith simply ignore the final cause and concentrate on the other causes. In effect, Bacon said that natural philosophers should stop asking, "Why?" and concentrate on asking "What?" and particularly "How?"

To justify natural philosophers abandoning 'Why?' as a question, Bacon held up a dazzling prospect: power. Power over nature and all the fruits and riches such power might bring. His programme might be paraphrased thus: stop agonising over why something happens and instead work out how it happens. Once you know that, then you will have power over it. For understanding the how of things, Bacon advocated experiment. He said, "[T]he secrets of nature betray themselves more readily when tormented by art than when left to their own course." This was a programme for the systematic experimental investigation of nature, with power over it as the prize.

Group Science

On 28 November 1660, The Royal Society was founded. Six years later, it was followed by the Académie des Sciences in France. The Royal Society was granted a royal charter by Charles II while thr Académie des Sciences was founded by Louis XIV. The prestige of natural philosophers was growing rapidly and the political establishment soon recognised the value of fostering and encouraging scientific investigation.

But the societies were also crucial in establishing the idea of science as a public enterprise, one in which scientists published their research, exposing their work to the scrutiny, and potential criticism of their peers. This was where the scientific and magical enterprises most clearly went their separate ways. For example, Francis Bacon was also an occultist, involved with various esoteric organisations including the Rosicrucians. But these organisations were deliberately secretive and their members, including Bacon, concealed their occult investigations rather than publishing them. Magic was something that men pursued in private; science was becoming increasingly public.

The difference lay in their basic methods. It might take a scientific genius to make a new discovery, but once that discovery was made it was possible for anyone to follow the reasoning behind it, understand it and apply it. Magic, as it was thought about then, was more akin to sport: something developed by adepts who might achieve mastery themselves but whose skills were personal, not public.

Bacon and other Renaissance figures pursued power over nature through both science and magic, but by the middle of the 17th century it was becoming clear that science was returning the better results. Curiously, however, one man remained strangely unaware of this, despite being the greatest scientist in history: Isaac Newton.

The Alchemical Scientist

The Scientific Revolution that Newton epitomised and, by his work, enabled, brought about a wholesale transformation in European thought. By its culmination, it would probably be fair to say that science had replaced not only the old ideas of magic, but also Christianity, as the main explanatory tool by which Europeans viewed their world. This had many related consequences. Abstract reasoning assumed a primary place in academic discourse. Nature, from being seen qualitatively, came to be analysed and explained quantitively. Natural philosophers stopped seeing the universe as an organism and started to describe it as a machine. And 'How?' became the key question that they asked, rather than 'Why?'.

Such wholesale changes constituted a fundamental redirection of Western civilisation and the way it thought about the world. But Newton, the man who did more to foster these changes than anyone else, himself remained a man of the old alliance of magic and science. While his scientific work brought him fame and renown, Newton himself spent much more time and effort on his alchemical and theological researches, his secrecy about these also being characteristic of the different approaches.

Newton's *Principia Mathematica*, published in 1687, was an astonishing synthesis of knowledge and a demonstration of the intellectual power that Francis Bacon had proclaimed in his *Novum Organum*. By his laws of motion, Newton united the local and the universal, showing that the same law that caused an apple to fall to Earth held the planets in their orbits.

Confirmation of Newton's theories was finally provided when his prediction that the Earth bulged slightly at the equator was confirmed by maritime scientific expeditions in the 18th century. Newton's laws, and the scientific methods by which they were arrived at, appeared able to explain everything - and for centuries, they did. Newly confident of the power of reason, and wielding its mathematical and scientific tools, scientists and scholars proclaimed a new Enlightenment and proceeded to remake the world. The Scientific Revolution appeared to be complete.

Antoine Lavoisier shown with his wife Marie-Anne, who was also a chemist

ALCHEMY AND CHEMISTRY IN AN AGE OF TRANSITION

THE ALCHEMISTS WERE OBSESSED WITH TRANSMUTING BASE METALS INTO GOLD. THOUGH THEY FAILED, THEIR INVESTIGATIONS HELPED GIVE BIRTH TO MODERN CHEMISTRY

WRITTEN BY **MARC DESANTIS**

Alchemy in late Medieval Europe brought with it many risks. The Roman Catholic Church was deeply opposed to its practice, being highly suspicious of its ancient connections to pagan practices and magic. Bodily immortality granted by potions such as the fabled 'elixir of life' had no place in a religious environment in which eternal life could be obtained only through the Christian sacraments. So noxious was alchemy during this time that both political and religious figures forbade its practice, and the Italian poet Dante Alighieri confidently stuck alchemists in Hell in the *Divine Comedy.*

Late Medieval alchemy also suffered from its continued lack of success in transforming lead into gold, one of its most prominent endeavours. Doubt as to the possibility of transmutation thus increased, though some alchemists continued doggedly in their quest.

The transition from alchemy, with its obsessions with the four elements as propounded by Aristotle, the search for the elixir of life, and the transmutation of base metals to gold to modern chemistry in the 17th century was not a smooth one. This was a hybrid period in the development of chemistry, one in which practitioners adhering to older alchemical precepts lived and worked under the same sky as those observing new, experiment-based principles.

Some figures who loom large in the history of science, such as Robert Boyle, or that paragon of rationality, Isaac Newton, actually straddled the divide between two kinds of chemistry. The discipline was not unique in this regard. The preeminent astronomers Johannes Kepler and Galileo Galilei both also devised astrological horoscopes, for example. Chemistry would at last find its recognisable, modern shape in the early 18th century, particularly in the person of the French chemist Antoine Lavoisier.

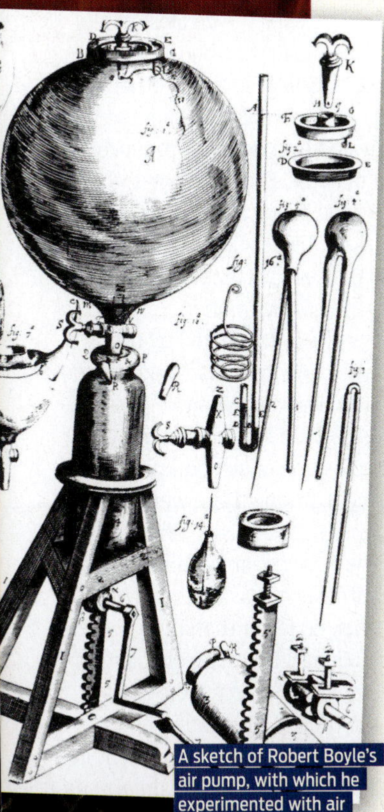
A sketch of Robert Boyle's air pump, with which he experimented with air

Robert Boyle

Robert Boyle (1627-1691) came from prosperous stock, being a scion of a landholding English family in Ireland that had colonised the island in the 16th century. The vicissitudes of fortune saw Boyle's once-prosperous father tossed into gaol for corruption. His father was later acquitted of the charges brought against him, subsequently bought more land in Ireland, and again grew rich, becoming in time the Earl of Cork.

Boyle's scientific career would be boosted by his family's wealth. He was tutored privately, and then sent to school at Eton.

The Englishman Robert Boyle, author of *The Sceptical Chymist*, did much to lead chemistry away from alchemy to a more scientific form

He could also afford to travel across Europe, including Italy, where he familiarised himself with Galileo's scientific contributions. Boyle received an honorary M.D. from Oxford, and also became a member of the Royal Society, a prominent British science association.

Boyle was thus no fringe figure, but a scientist standing very much in the centre of scientific learning of the era. He was in contact, via letters, with other scientists, exchanging news of his own discoveries with them and disseminating theirs to wider audiences. He was also a powerhouse of chemical inquiry, running his own private laboratory, overseeing the experiments of others, and offering funds to yet more to conduct their own research. His 1661 work, *The Sceptical Chymist*, did much to establish chemistry as being something that stood apart from earlier alchemy, it being based squarely on experimentation, and not old-fashioned appeals to authority.

Among Boyle's achievements was his pioneering work on analytical chemistry, in which he enthusiastically performed analysis via solution chemistry instead of the traditional pyrolysis, in which substances were burned. Fire, Boyle realised, could not break materials back down into their constituent ingredients, and thus pyrolysis could never be a useful analytical tool. Solution chemistry was much more reliable. Boyle used acids and alkalis - reagents - that could actually dissolve substances so that they could be studied properly.

Boyle himself may be judged one of the first true chemists. He quantified what he did, and took careful measurements. One of his greatest achievements, concerning air pressure, and today known as Boyle's Law, holds that the volume of a given mass of air is inversely proportional to the pressure applied to it.

Boyle also rejected the ancient theories of chemical composition that held that matter was composed of earth, air, fire, and water. He had no time for either Aristotle's theory of the Four Elements nor the Three Principles theory, which saw matter as comprised of sulphur, mercury, and salt. Boyle was unsatisfied with the prevalent notion that all matter was a combination of four elements, but given the limitations of his day, he had no notion as what these might actually be nor how many of them there were. And despite his sterling work shepherding chemistry into the modern form it holds today, Boyle still believed that alchemical transmutation might be possible. At his death in 1691, he left another scientific giant of the age, Isaac Newton, with a bit of red earth that Boyle thought might be capable of transforming mercury into gold.

ALCHEMIST JOHANN FRIEDRICH BÖTTGER WAS IMPRISONED UNTIL HE WORKED OUT HOW TO CHEMICALLY PRODUCE PORCELAIN IN 1708

Isaac Newton: The Last of the Magicians

Though Isaac Newton (1642-1727) is today best-known for his ground-breaking achievements in physics, he was also an enthusiastic investigator of chemical science, filling his notebooks with numerous entries on various substances such as mercury and antimony. Newton devoted much time to alchemy, purchasing several alchemical texts, as well as the requisite apparatus. Alchemy had already become a looked-down-upon endeavour, so Newton made use of unpublished works that passed from one alchemist to another on an informal basis. Newton himself made his own handwritten copies of some of the alchemical works that came into his possession.

Dubbed the 'last of the magicians', Isaac Newton was both a modern scientist and an avid practitioner of alchemy

Physics and alchemy differed markedly in their philosophical approaches. Physics, as practiced by Newton, had no place for 'spirit' in the operation of nature. Every act was purely physical. Alchemy, however, was deeply spiritual in its outlook, seeing nature as being suffused by spirit. And unlike matter as it was described in physics, which was lifeless, alchemy saw matter as being filled with active principles that were the prime causes of natural phenomena.

Newton was no dabbler in the pseudoscience of alchemy, despite his reputation as perhaps the greatest of all scientists. The time spent on the subject was extensive. Unlike so many other, earlier, alchemists, Newton does not seem to have been motivated by a desire to produce gold from base material. From his notes, this appears to have been of only limited interest to him. Instead, his real aim was the truth. Newton was a searcher, and alchemy was another avenue of exploration.

Newton was also fascinated by the Rosicrucians, a 17th century brotherhood of mystical alchemists. When he passed, his library was found to contain an extensive collection of alchemical works, and prominent among these were books concerning the Rosicrucians. These included a copy of the Rosicrucian Manifesto that is replete with Newton's own handwritten notes. Further, no fewer than nine volumes penned by Michael Maier, a noted authority on the Rosicrucian Society, were in his library at the time of his death.

The extent of Newton's involvement in alchemical research would remain obscured

Antoine Lavoisier conducts an experiment with his solar furnace

for centuries after his death. In the 1930s, Newton's papers appeared on the auction block, and some were purchased by the famed economist John Maynard Keynes. From these and others of Newton's papers that he collected, Keynes discovered a side of Newton that was previously scarcely known to history. Keynes concluded that "Newton was not the first of the age of reason. He was the last of the magicians, the last of the Babylonians and Sumerians, the last great mind which looked out on the visible world with the same eyes as those who began to build our intellectual inheritance rather less than 10,000 years ago."

Chemistry in the Age of Reason

The 18th century would see the emergence of modern approaches to science. Discoveries were shared freely by scientists with their peers, not kept secret, and scientific societies were founded with government support across Europe.

The 'Chemical Revolution' transformed the discipline of chemistry in the 18th century. In a sharp break with the past, chemistry would now be characterised by precision in measurements and rigidly logical theory. One of the Revolution's foremost, and earliest, advances was to identify different chemicals as distinct materials with their own unique properties. Speculation would have to be supported by data derived from verifiable, repeatable experiments.

The spirit of the new age was exemplified by the Frenchman Antoine Lavoisier (1743-1794). His interest in chemistry was kindled while he was still in law school. Though he would pass the bar examination, his heart would belong to chemistry. Lavoisier, who had by then already contributed research into better street lighting, became a member of France's Academy of Science when he was just 21 years old.

Lavoisier would go on to make many other contributions to chemistry. He demonstrated by experiment that, contrary to hoary alchemical belief, water did not transmute into earth when heated for a long duration. This experiment worked to establish the principle of mass conservation, one of chemistry's bedrock tenets.

In 1775, Lavoisier identified oxygen, a unique gas that was especially supportive of combustion, more so that ordinary air. On the basis of the new-found oxygen, he would also do away with another outmoded alchemical theory, that of the oily, fire-starting substance phlogiston. Lavoisier would also establish the chemical elements as those substances that could not be reduced to simpler substances by chemical means. Chemicals were either elements or compounds of elements.

He also helped name chemicals, thereby originating the systematic nomenclature used today in modern chemistry. By Lavoisier's death at the age of 51 in 1794, chemistry had acquired the contours that it still holds today.

ANGELO SALA

The chemist Angelo Sala (1576-1637) stood apart from the mass of his fellow Italians on account of his Calvinist Protestant religion. Like Boyle and Newton, he straddled the dividing line between alchemist and chemist, being a believer in the theory of Three Principles. One of his greatest contributions was proving that substances retained their identities even after they had been combined with others. This theory was by no means new, but it ran contrary to Aristotle, who believed that substances' individual identities were destroyed utterly when they reacted with one another, leaving behind no traces of the original combining materials.

Sala thought otherwise, and sought the truth. In 1617, in his book *The Anatomy of Vitriol*, he informed the world of an experiment he had conducted concerning chemical structure. First, he dissolved a known weight of copper in heated sulphuric acid. He then added water, resulting in 'blue vitriol' - copper sulphate hydrate. Next Sala converted the substance to copper oxide, then reduced it back into copper. Upon weighing the copper, it matched the amount he had started the experiment with. The copper had not been lost at all, but kept its identity even after combination, and this was proven by its complete recovery. Sala had thereby shown that Aristotle's age-old Four Elements theory was incorrect.

Sala's also demonstrated that his self-made blue vitriol was identical with that of naturally-occurring blue vitriol. This crushed the entire theory of the transmutation. Traditional alchemical thought had held that minerals were living things, with souls, that grew in the earth. Sala could make synthetic blue vitriol with a run-of-the-mill chemical reaction, thereby proving that it was not a living thing, and that the ancient belief in transmutation was invalid.

17th century Italian chemist Angelo Sala disproved Aristotle's Four Elements theory

CHARLES DARWIN

IN A TIME WHEN THE CHURCH RULED THE ROOST, DARWIN'S THEORY OF EVOLUTION SHOOK VICTORIAN SOCIETY TO ITS CORE

WRITTEN BY LAURA MEARS

Science and religion aren't usually known for going hand in hand, and Charles Darwin's revolutionary ideas about evolution widened the divide between the two, sparking controversy at a time when religion was paramount in people's lives. Renowned for his theory of evolution, Darwin's scientific beliefs contradicted the Victorian belief in God's divine creation. While many clergymen branded Darwin a blasphemer, however, his ideas were quickly accepted by many, going on to form the basis of natural science as it is today.

Born on 12 February 1808, Charles Darwin was raised in a wealthy, liberal-minded family. Despite being brought up with Christian morals and teachings, he was encouraged to push himself and explore his own ideas. With two grandfathers of significance, it was little wonder that the young Darwin was so curious.

His paternal grandfather, Erasmus Darwin, who had died several years before the young Darwin's birth, had been a physician of great reputation - he was offered the position of Royal Physician, but turned it down. What made Erasmus so notorious, however, were his ideas about transmutation - he thought animals had an inbuilt capacity for change - which preceded Charles Darwin's similar theory of evolution. However, Erasmus's suggestion of transmutation wasn't received well and he was criticised for denouncing his creator, God.

Intending to follow in his father and grandfather's footsteps, Charles Darwin decided to pursue medicine at university in 1825 and moved to Edinburgh to study. However, after watching surgeries performed without any kind of sedative, he soon realised that the discipline wasn't for him and his sensitive stomach. Despite this, Edinburgh turned out to be the ideal place for Darwin to cultivate his liberal ideas; the city was filled with radicals who debated outrageous and revolutionary theories that would never be tolerated in Oxford or Cambridge.

The open-mindedness of Edinburgh couldn't hold Darwin, however, and in 1827 he moved to Cambridge University to study divinity, with the intention of becoming a clergyman. While he wasn't particularly passionate about it, his course enabled him to pursue his real love - collecting beetles.

Upon graduating from Cambridge in 1831, Darwin intended to find himself a position within the Church, but another opportunity arose, which would give him a chance to sate his ever-growing curiosity about life. He was invited to join the ship HMS Beagle as the ship's 'gentleman naturalist' as it went on a two-year voyage around the world. However, what was meant to be two years eventually became five, and Darwin visited four continents over this time.

Yet this chance of a lifetime came at a cost for the young Darwin, who suffered terribly from seasickness. For the first few weeks of setting sail, it's said that the only food that Darwin could eat was raisins - the only thing his stomach could handle. Unfortunately, this was the first illness of many in Darwin's adult life, leaving him debilitated for long periods. One modern theory is that during his travels, Darwin caught a tropical fever that tormented him until his death in 1882.

His illnesses never hindered his research, however, and Darwin continued to collect samples of animals on his travels. Perhaps the most notable visit was his time spent aboard the HMS Beagle is the trip to the remote Galápagos Islands in the Pacific Ocean, where the ship moored for over a month. On these islands

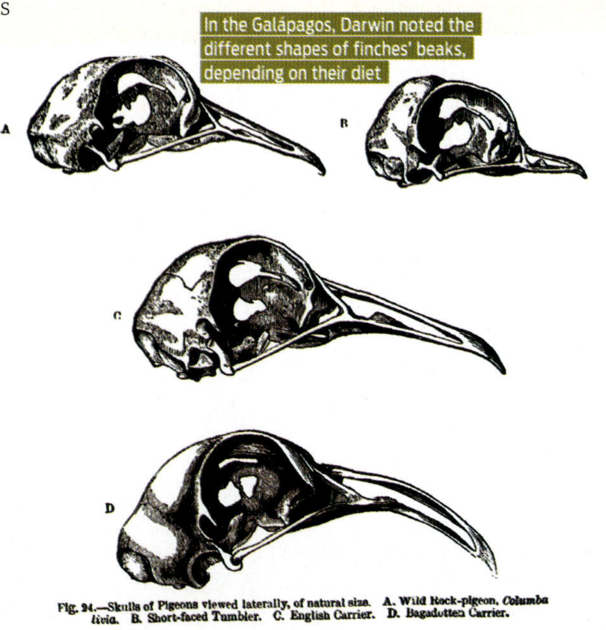

In the Galápagos, Darwin noted the different shapes of finches' beaks, depending on their diet

Images source: Adobe Stock, Wiki

27th December, 1831
Plymouth, England
Darwin agrees to take the position of ship's captain's assistant but has to wait three months before the Beagle sets sail. The first week of the voyage would prove to be trying as he was confined to his bed with seasickness.
15th September - 20th October 1835
Galapagos Islands
Darwin was most fascinated by the enormous tortoises and iguanas that populated the islands. He wouldn't realise until later when he returned to England that the different species he saw were actually specific to each of the islands.
St Jago
16th Januar 1832
Darwin overcomes his seasickness at the Cape Verde islands and realises what a fantastic opportunity the Beagle's voyage represented. His diary is filled with excitement as he sees exotic vegetation and animals for the first time.
January-February 1835
Salvador, Brazil
29th February, 1832
The Beagle makes anchor in South America for the first time and Darwin sets his eyes on the rainforest. The sheer vastness of his mission was dawning upon him as he loaded up the Beagle with samples to be sent back to England.
Chiloé, Chile
Darwin's ideas expanding on Lyell's theory of an Earth in a constant state of movement were confirmed when he witnessed the massive eruption of Mount Osomo. It was a tremendous and moving sight.
3rd August, 1833
Darwin spent some of his happiest days exploring the wildlife of Argentina. At Rio Negro he rode with the gauchos, hunting for dinner and avoiding rebel forces. He was impressed by General de Rosas, who gave permission for Darwin's passage.
Rio Negro, Argentina

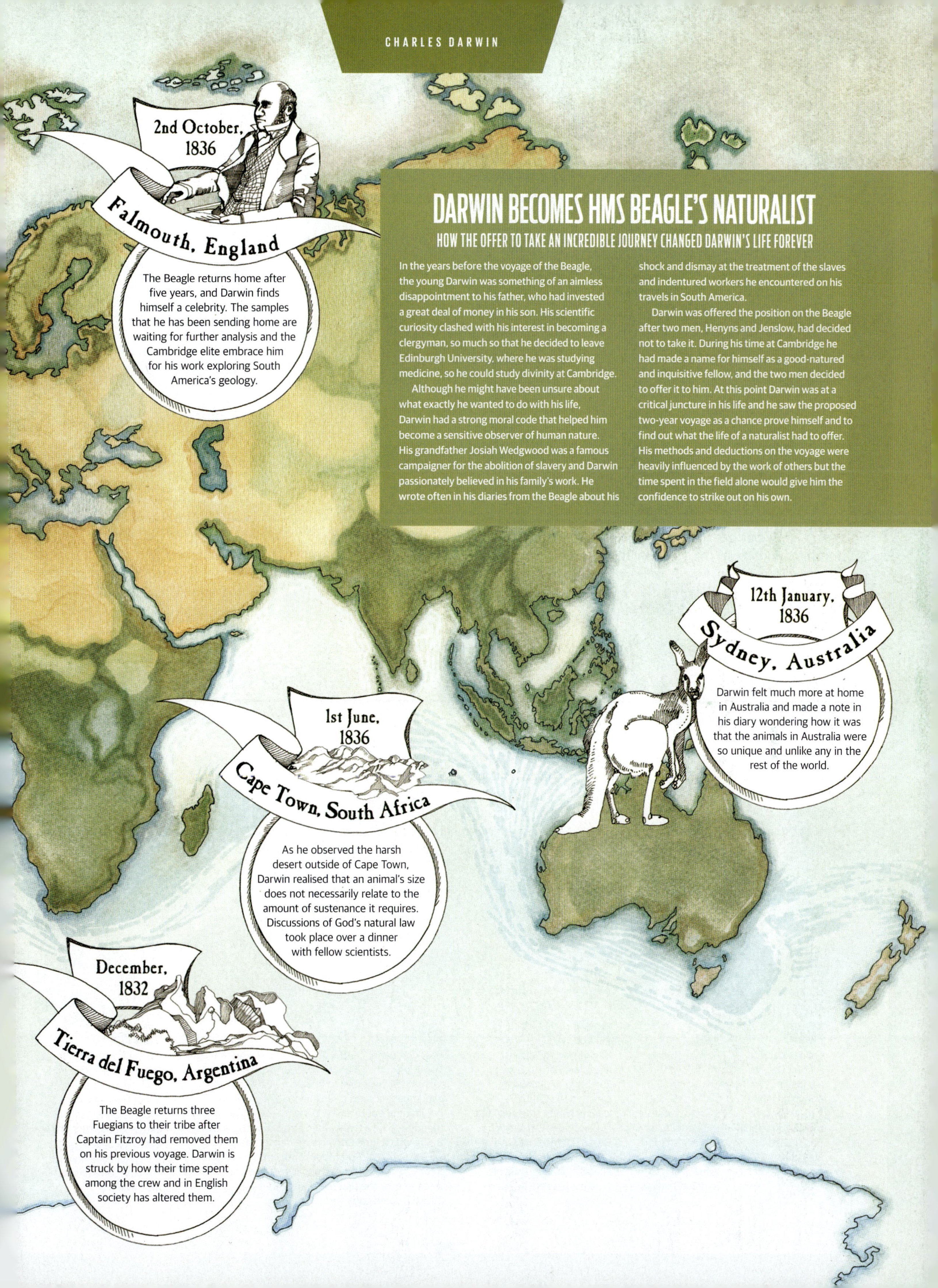

DARWIN BECOMES HMS BEAGLE'S NATURALIST

HOW THE OFFER TO TAKE AN INCREDIBLE JOURNEY CHANGED DARWIN'S LIFE FOREVER

In the years before the voyage of the Beagle, the young Darwin was something of an aimless disappointment to his father, who had invested a great deal of money in his son. His scientific curiosity clashed with his interest in becoming a clergyman, so much so that he decided to leave Edinburgh University, where he was studying medicine, so he could study divinity at Cambridge.

Although he might have been unsure about what exactly he wanted to do with his life, Darwin had a strong moral code that helped him become a sensitive observer of human nature. His grandfather Josiah Wedgwood was a famous campaigner for the abolition of slavery and Darwin passionately believed in his family's work. He wrote often in his diaries from the Beagle about his shock and dismay at the treatment of the slaves and indentured workers he encountered on his travels in South America.

Darwin was offered the position on the Beagle after two men, Henyns and Jenslow, had decided not to take it. During his time at Cambridge he had made a name for himself as a good-natured and inquisitive fellow, and the two men decided to offer it to him. At this point Darwin was at a critical juncture in his life and he saw the proposed two-year voyage as a chance prove himself and to find out what the life of a naturalist had to offer. His methods and deductions on the voyage were heavily influenced by the work of others but the time spent in the field alone would give him the confidence to strike out on his own.

At the time, Darwin was ridiculed by many, particularly for his idea that humans were descended from apes

Darwin's grandfather, Erasmus Darwin, was ostracised after theorising about transmutation between species

(which were teeming with wildlife), Darwin studied finches, mockingbirds and tortoises. He researched these animals thoroughly and noted that different finches' beaks were shaped differently depending on the food that they ate, marking his first serious foray into his ideas on natural selection.

As soon as he returned home in 1836, Darwin found himself mulling over what he'd experienced during his time away. He published an account of his travels, but his mind was preoccupied; he was dwelling on the early formation of his theory of evolution. His feelings were conflicted, he was troubled that his findings contradicted his Christian values and feared being ostracised and condemned by society for his controversial ideas. For this reason, Darwin didn't publish his theory when he first came up with it.

In order to satisfy his mind, however, Darwin decided to keep studying the specimens that he'd collected during his travels, and continued researching his idea so that he could gather together enough evidence to prove his point. By 1838 he began drafting his ideas, forming the start of *On the Origin of Species*. His research remained personal; Darwin's fears meant that only he and his close friends knew about his theory.

Science wasn't the only thing on Darwin's mind upon his return, however. On 29 January 1839, he married his first cousin, Emma Wedgwood. It's clear that it wasn't particularly a priority for the scientist - he famously wrote a pros and cons list in order to decide whether to marry at all, with some of the pros charmingly stating that a wife is "better than a dog" and provides "charms of music and female chit chat", while the cons included "forced to visit relatives" and "less money for books".

His relationship with Emma Wedgwood produced ten children. Tragically, three of their offspring died, and it was the death of his eldest daughter, Anne, at the tender age of ten, that struck him the most. His children were frequently unwell, and when Anne died, he began to investigate the detriments of genetic inbreeding.

Soon, however, Darwin's mind was forced away from his family and brought back to science. In 1858 Darwin received a letter from a long-time admirer named Alfred Russel Wallace. Inspired by Darwin's travels across the world on HMS Beagle, Wallace had embarked on his own voyage and had come to the same conclusions about natural

TIMELINE

1808
Charles Darwin is born. Despite a Christian upbringing, his family are open-minded. Erasmus Darwin had the idea of species transmutation.

1825
Darwin studies medicine at Edinburgh University, but quickly realises the brutality of surgery isn't for him. Edinburgh is full of like-minded thinkers.

1827
Having decided against medicine, Darwin leaves Edinburgh to pursue a career in the Church, leading him to study divinity at Cambridge University. He isn't particularly fond of religion, but his time here enables him to pursue his greatest passion: beetle collecting.

1831
After graduating in 1831, Darwin is intent on finding a position in the Church. However, before he can pursue this, he is invited aboard HMS Beagle as a 'gentleman naturalist'. Unable to refuse this once-in-a-lifetime opportunity, Darwin accepts. What is intended to be a two-year trip ends up lasting five, during which time Darwin experiences the local wildlife and geology of four continents. He suffers greatly from seasickness, and it's said he only ate raisins for the first few weeks, as it was all his stomach could cope with.

1836
Upon his return to Britain, Darwin decides to hold back on going public about his ideas about evolution. Instead, he gathers more supportive evidence by studying the specimens that he has collected during his travels.

selection as Darwin. He sought his hero's advice on how to publish his findings.

Considering that he hadn't gone public with his ideas yet, Darwin was distraught. He'd come to these conclusions first, and he didn't want Wallace to get all the credit for the idea. However, he recognised that Wallace's research was valid, and he didn't want to undermine this. Wallace was still abroad and uncontactable, so Darwin was at a loss of what to do morally.

Eventually, Darwin decided to go public with his ideas. To overcome the issue with Wallace, Darwin presented their ideas alongside each other. Their ideas on evolution and natural selection were presented to the Linnean Society, the leading natural history body in Britain. When he returned to Britain, Wallace agreed that Darwin had acted fairly.

Despite the fact that both Wallace and Darwin were considered co-discoverers of natural selection, Darwin went on to eclipse Wallace thanks to the publication of Darwin's *On the Origin of Species* in 1859. It was this literary achievement that captured the imagination of the public, and it was an incredible success. *On the Origin of Species* went on to become a bestseller, and it was translated into many languages. Yet not everyone was quite so accepting of Darwin's book. Many members of the Church accused Darwin of blaspheming, as the book contradicted the idea of divine creation that was written in the Bible. However, some religious figures interpreted Darwin's theory as an instrument of God's design. It wasn't just within religious and scientific circles that Darwin's text sparked controversy - his theory became part of popular culture, and Darwin was parodied, ridiculed and caricatured in many newspapers at the time, particularly for having hinted at the idea that humans were descended from apes.

In June of 1860, a debate was held about evolution at Oxford University, where Darwin's biggest supporters went head to head with religious leaders. In what is often seen as a turning point in the relationship between religion and science, Bishop Samuel Wilberforce taunted Thomas Huxley - one of Darwin's close friends and a staunch believer in evolution - about his ape ancestry, and Huxley retorted sarcastically. Both left feeling that they'd won the debate. Regardless of who won, it's a prime example of how much Darwin's theory shook Victorian society.

Despite his early fears, Darwin had borne the brunt of ridicule and disbelief at his revolutionary theory and was no longer a stranger to criticism. In 1871, overcoming his doubt, his discourse on how humans were descended from apes turned explicit in his latest work, *The Descent of Man*. While many Victorians were divided about the idea of evolution, his ideas were gaining credence, and many notable figures at the time were converting to a Darwinist mindset.

As his health deteriorated, Darwin became a virtual recluse at his rural home in Downe, Kent, where he was nursed by his wife and children. He seldom received visitors, but he never let his illness impact his work.

On 19 April 1882 Darwin passed away, having expressed a wish to his wife Emma that he be buried in a local graveyard, despite his religious stance (as an agnostic). His close friends, however, had other ideas, and he was eventually laid to rest at Westminster Abbey.

THE IMPACT OF INBREEDING

WHY DARWIN'S MARRIAGE TO HIS RELATIVE INFLUENCED HIS IDEAS

Darwin and Emma went on to have ten children. Tragically, three died in childhood - a son and a daughter during their infancies, and another daughter, Anne, at the age of ten.

The death of Anne had a huge impact on Darwin and he began researching the effects of inbreeding. Citing the negative impact of self-fertilisation on the life and development of orchids, Darwin questioned whether his own marriage and relationship to Emma were causing illness and weakness in his children.

Such was his worry that he sought to change the laws on marriage between cousins. At the time of the 1871 census, he lobbied to add questions surrounding the issue, but he was refused outright. After all, in questioning the morality of marrying a cousin, Darwin was challenging the marriage of Queen Victoria, who had herself married a cousin - Prince Albert.

Fortunately, Darwin's remaining seven children lived long, fulfilling lives, each having inherited Darwin's own insatiable curiosity and intelligence. Three of his sons were eventually knighted for their services to astronomy, botany and civil engineering respectively.

1839
After using pen and paper to weigh up the pros and cons of marrying his cousin Emma, he decides that companionship overrides his need for intelligent conversation.

1858
In the summer of 1858, Darwin receives a letter from an admirer, Alfred Russel Wallace, seeking advice on how to publish his independent findings of natural selection, which prompts Darwin to go public with his ideas.

1859
Fearing Wallace will publish before him, Darwin goes public with his findings. In 1858 he presents his ideas to the Linnean Society alongside Wallace's, crediting him too. In 1859 Darwin publishes *On the Origin of Species*, which is a bestseller. The book sparks controversy in religious circles and some call Darwin a blasphemer. Despite this, many come to agree with Darwin's ideas.

1860
Darwin's biggest believers go head to head with Bishop Samuel Wilberforce in a debate at a meeting for the British Association For The Advancement Of Science. Held at Oxford University, both sides depart feeling that they've come out as winners.

1869
As a response to critics, Darwin edited *On the Origin of Species* to strengthen his arguments. By the fifth edition, Darwin quoted the phrase 'survival of the fittest' from economist Herbert Spencer, which has often been incorrectly attributed to Darwin. The phrase was more fitting for Darwin's ideas.

1871
Having previously shied away from explaining human evolution, Darwin plucks up the courage to explicitly state that humans are descended from apes in his latest publication, *The Descent Of Man*.

1882
Having been wracked with ill health, Darwin passes away in the company of his wife and a few close friends. Despite expressing a desire to be buried in a local graveyard, he is laid to rest at Westminster Abbey.

EDISON VS TESLA

SPARKS FLEW WHEN THESE TWO ELECTRICITY HEAVYWEIGHTS CLASHED OVER THE FUTURE OF ENERGY IN A FEUD THAT CHANGED THE WORLD

WRITTEN BY **LAURA MEARS**

"AS PANIC INTENSIFIED, EDISON STARTED TO SPEAK UP, LAYING THE BLAME FOR THE DEATHS FIRMLY AT HIS RIVAL'S DOOR"

The world's first Westinghouse AC generator, at Ames Hydroelectric Plant, Colorado

Images source: Alamy, Adobe Stock, Getty

The large dog trembled as it was led onto the stage. A black Newfoundland, muzzled, with electrodes trailing from his limbs - one at the front, another to the rear. The cage door creaked shut, and the crowd waited. Harold P Brown threaded the wires into a generator, and flipped the switch. The dog yelped and stiffened as Brown turned the dial. He increased the voltage until 1,000 volts of direct current were coursing through the animal's body. But when the switch was flipped again, the poor creature was still alive.

"We shall make him feel better," Brown announced, taking the wires and connecting them to a different generator at the front of the hall. This time, the prescribed dose was 300 volts of alternating current. The switch was flicked, the dog twitched and then it died. "Alternating current is suitable only for the dog pound, the slaughterhouse and the state prison," he declared, triumphantly.

The year was 1888. The War of Currents was under way, and the American inventor was trying to prove a point to his shocked audience. The country needed electricity, but how it would be delivered was a matter of furious debate. At the centre of the furore were two of the greatest inventors of the time: Thomas Edison, pushing direct current, and the Serbian Nikola Tesla, advocating for alternating current.

Edison was an all-American inventor and entrepreneur. He had begun his career in electricity, rising from lowly beginnings as a

The Westinghouse AC generator, at Ames Hydroelectric Plant, Colorado

telegraph operator to become the esteemed inventor of the commercial light bulb and the phonograph. He was meticulous, methodical, and had a natural head for business, and the Edison Electric Light Company was supplying the current that lit people's homes around the United States of America.

His entire empire was built upon direct current - electricity flowing in one direction - and his company had been working to roll out their generators to supply the nation with the juice that they needed to keep the lights on and the music playing. But Nikola Tesla had a different idea.

The Serbian inventor had arrived in the USA in 1884, and was originally employed by Edison himself. He worked at Edison Machine Works for the princely sum of $18 a week, and under the instruction of the master inventor, was charged with the task of finding a better way to transmit electricity over long distances.

People needed about 100 volts to power their electric lights at home, so Edison's electricity was delivered at this fixed value in thick copper cables. But, as electricity courses through wires, some of the energy escapes. This meant building lots of small generators close to people's houses so that power wasn't lost along the way - a costly and inconvenient solution.

Tesla's big idea was to use alternating current on the lines. Instead of travelling just one way, it flipped back and forth, changing direction several times a second. This creates varying magnetic fields, which can be used to create transformers. These devices can be used to increase or decrease the voltage on demand. The thought was that electricity would travel over long distances at high voltage, and would then be decreased to a lower voltage when it reached people's homes. This wasn't possible with direct current, however.

Edison was less than impressed. A practical man, he thought Tesla was a dreamer, and a 'poet of science'. "His ideas are magnificent, but utterly impractical," he said.

In Edison's mind, alternating current was dangerous. It flipped back and forth in such a way that it could interfere with the heart, and high-voltage power lines above the street were just unthinkable. Besides, Edison was a prolific inventor, with more than 1,000 patented inventions in the US alone. People wouldn't use his inventions if alternating current was adopted. He refused to support Tesla's ideas, offered him a $7-a-week pay rise for his trouble, and his apprentice walked out of the door.

So annoyed was Tesla at Edison's lack of foresight that he set out to bring his alternating current inventions to the USA on his own. After a good start, he fell upon hard times, and with barely enough money to live, he took a job as a manual labourer, earning just a few dollars a week digging ditches. Eventually, he scraped together the much-needed funds to set up his own firm, the Tesla Electric Company.

His innovative ideas caught the attention of wealthy railway tycoon George Westinghouse. Fat on the profits of a booming transport business, Westinghouse was ready for a new challenge, and taking on the United States' emerging electricity giants seemed like just the thing. He'd dabbled in alternating current before, and Tesla's inventions were exciting.

A year before the execution of the poor Newfoundland, Tesla had filed seven patents to protect his new ideas. He devised a complete power system, designed not only to generate electricity but also to carry it over long distances, increase or decrease the voltage, and he even created motors and lights that could be used at the other end.

Tesla was no businessman, he was a pure inventor, but Westinghouse was the Steve Jobs to his Steve Wozniak, and he took Tesla's ideas and spun them into a business that could take Edison down. Tesla sold his patents to Westinghouse for $60,000, taking $10,000 in cash, 150 shares, and an agreement that he'd make $2.50 for every horsepower of electricity that Westinghouse managed to sell. This was the beginning.

Westinghouse Electric began setting up generators across the US, reaching into areas

Thomas Edison experiments in a laboratory, circa 1900. He was known for his meticulous approach to problems

that Edison's generators couldn't access, and undercutting them in inner cities to poach their customers. The price of copper was going up, Edison's business was under threat, and to top it off, his inventions were being ripped off by his competitors. "Westinghouse will kill a customer within six months after he puts in a system of any size," Edison desperately fumed.

In 1888, his company issued a written warning to journalists detailing the dangers of alternating current, and his associate, Harold P Brown, began his awful public campaign of animal electrocutions. After the incident with the dog, he took his twisted show on tour around New York City, dispatching stray animals in front of the

The World's Columbian Exposition, lit by Tesla's alternating current

EDISON'S WEIRDEST INVENTIONS

THOMAS EDISON WAS A PROLIFIC INVENTOR, WITH HUNDREDS OF BIZARRE CONTRAPTIONS TO HIS NAME

DOLL VOICES

The inventor miniaturised his famous phonograph, encased it in tin, and put it inside a doll to create a voice box. The sound quality was poor, and the resulting noises reportedly frightened customers.

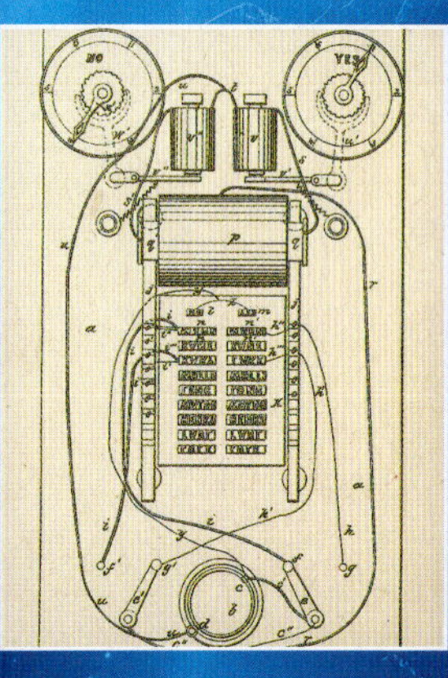

VOTE RECORDER

Edison's first invention was designed to count votes. Voters flicked a switch to indicate their choice, and an electrical current travelled to a machine, which recorded 'yes' or 'no' on dials.

MICROSCOPIC VIDEO

'Peephole kinetoscopes' combined light bulbs with photographs to create moving pictures, viewed through a microscope. The four-foot high contraptions only allowed one person to watch at a time.

TESLA'S STRANGEST IDEAS

NIKOLA TESLA WAS A BIG THINKER, WITH IDEAS EXTENDING WAY BEYOND HIS TIME

LONG-RANGE WIRELESS POWER

Tesla thought he could transmit electrical power wirelessly from Wardenclyffe Tower in Long Island. But he ran out of money, and the project was abandoned.

ROBOTIC WORKFORCE

With aspirations to create labour-saving robots that would do menial tasks for humans, Tesla developed a remote-controlled boat. However, his robot race has yet to be realised.

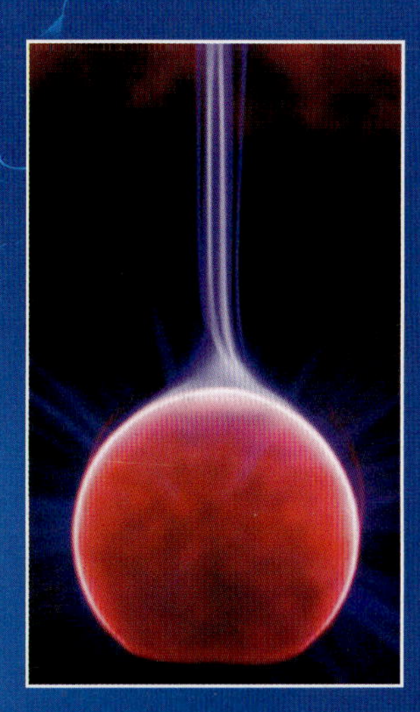

DEATH RAY

The inventor is reported to have designed a 'death ray' that created a beam of particles that could bring down aircraft. His papers on it have never been found.

horrified eyes of journalists and denouncing Westinghouse and Tesla's alternating current as a menace to public safety.

Westinghouse was enraged, and wrote a letter to Edison berating him for the actions of his associates. "I believe there has been a systemic attempt on the part of some people to do a great deal of mischief and create as great a difference as possible between the Edison Company and The Westinghouse Electric Co." Tesla also retaliated, setting up public demonstrations of his own, in which he let 250,000 volts of alternating current course through his own body to the astonishment of onlooking crowds. As the saying goes, "It's the volts that jolts, but the mills that kills." Tesla was completely sure of his technology, even if Edison wasn't.

Relations deteriorated, and the rivalry escalated. In 1889, the 'electric wire panic' began. Several linemen working on alternating current power cables had died in the line of duty, prompting widespread public anxiety about the installation of these high-voltage lines. The deaths added fuel to Edison's fury about the safety of Tesla's technology. Until this point, he'd remained behind the scenes, allowing Brown and the Edison Electric Light Company to take the lead in the campaign against Westinghouse, but as panic intensified, Edison started to speak up, laying the blame for the deaths firmly at his rival's door.

'Topsy' the elephant was electrocuted in 1903 with alternating current

Things were about to get worse. William Kemmler was a peddler from Buffalo, New York. A heavy drinker, and a bully, he had killed his partner with a hatchet in a fit of rage, and in 1890, he was awaiting execution. Edison had been approached to advise on a more humane method of dispatching criminals after a spate of failed hangings, and although opposed to the death penalty in principle, he had eventually responded with a recommendation. So sure was he about the dangers of alternating current that he suggested a Westinghouse generator.

Westinghouse fiercely opposed the idea. He thought it was unnecessarily cruel, and instructed his lawyers to mount an appeal for Kemmler's life, putting $100,000 behind the cause. The moniker 'executioner's current' was not one he wanted attached to his company. Edison cornered his opponent, and responded by accusing Westinghouse of prioritising his commercial reputation over the welfare of the convict, arguing that alternating current would be a swift and

effective method of execution. Behind his competitor's back, Edison's company helped Brown to source second-hand Westinghouse motors to supply the electric chair, but the execution didn't quite go to plan. When Kemmler was strapped in, a whopping 1,300 volts were passed over his body. The ordeal lasted 17 seconds, but he didn't die. The audience screamed as he regained consciousness and his clothes caught fire. Only when the device was ramped up to 2,000, and run for four minutes, did his body finally concede to the assault. It was a grisly show. "They would have done better using an axe," Westinghouse grimly observed.

The ugly back and forth between the USA's two electricity giants continued, until in 1893, a grand occasion presented the opportunity to end the feud once and for all. But by that point, Edison had dropped out of the race. Merger after merger had whittled the competing electricity companies down to a handful, and Edison's firm had been swallowed by Thomson-Houston to form General Electric. With it went his patents and his stake in the competition. The event to end all events went ahead in his absence, and was the final nail in the coffin of his empire.

The World's Columbian Exposition would be held to honour the 400th anniversary of Christopher Columbus's arrival in America. The event expected 27 million guests, and it needed to be lit. General Electric put in a bid to power the fair for $554,000, but Westinghouse undercut them, promising to keep the lights on for just $399,000. General Electric would need mountains of copper wire to transmit enough power, but Tesla's alternating current allowed Westinghouse to offer the same at a fraction of the cost.

The exposition glowed, and the requests for alternating current snowballed. Westinghouse was given access to Niagara Falls to use the torrents of water to generate hydroelectricity, and even General Electric switched to AC. The war was over, and Tesla's side had won.

Later, Edison went to see Tesla speak, and when his former apprentice noticed him, he asked the audience to give Edison a round of applause. Despite their differences, and the attacks on his work that Edison had sponsored, underwritten, and even directly engaged in, Tesla greatly admired Edison, and later said: "The effect that Edison produced upon me was rather extraordinary. I saw how this extraordinary man, who had had no training at all, did it all on his own."

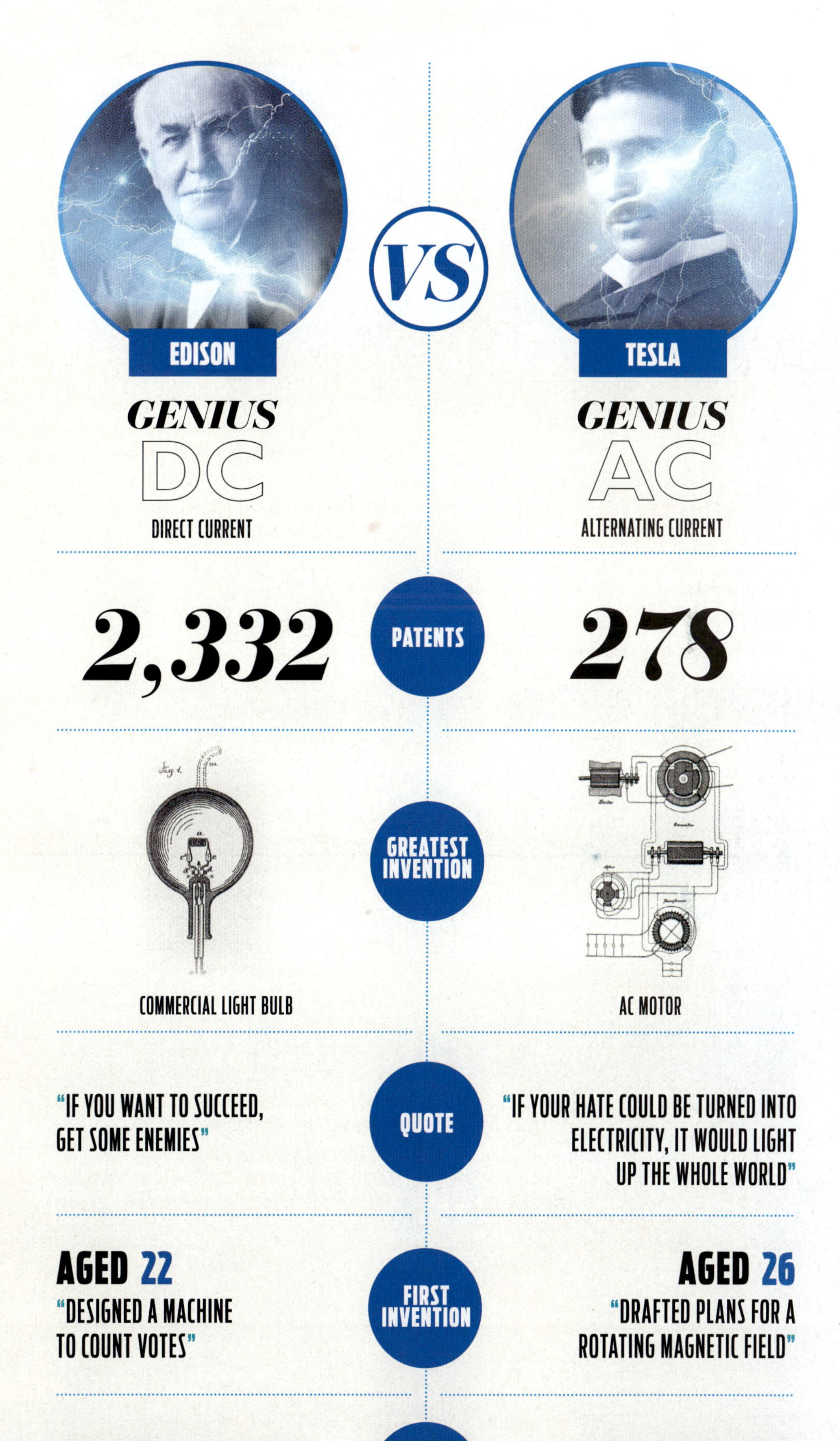

MARIE CURIE

A PIONEER IN NUCLEAR PHYSICS AND CHEMISTRY, MARIE CURIE MADE TREMENDOUS CONTRIBUTIONS TO MEDICAL SCIENCE AND OTHER FIELDS

She coined the term 'radioactivity'; she was the first woman to win the Nobel Prize; and then, just for good measure, she won it a second time.

Marie Curie explored the properties of radioactivity, and discovered two elements: radium and polonium. She applied her acquired knowledge to the field of medical science through the use of diagnostic x-rays and early assessments of its capacity to fight cancer. Her achievements were remarkable; however, they were even more noteworthy because she was a woman whose intellectual and scientific prowess were undeniable in a professional arena dominated by men.

Born in the city of Warsaw, then a part of Imperial Russia, Maria Salomea Sklodowska was the fifth and youngest child of Wladyslaw and Bronislawa Boguska Sklodowski, both well-known educators who prized the pursuit of academic excellence. Her mother died of tuberculosis when Maria was only ten years old. Polish nationalism was a hallmark of the family's worldview, and the loss of property experienced during support of such movements and periodic uprisings had left them struggling financially. Maria was educated in the local school system initially, while her father, a teacher of mathematics and physics, provided additional learning opportunities, particularly after the Russian authorities restricted laboratory instruction in the schools. Wladyslaw brought his equipment home and taught his children there.

A gifted student, Maria excelled in secondary school. However, she was prohibited from attending college because she was female. Along with her sister, Bronislawa, she enrolled in the 'Floating' or 'Flying' University, a clandestine college that conducted classes out of sight of the authorities, and also supported Polish nationalistic ideals. Prospects for higher education were more favourable to women in Western Europe, and the sisters came to an agreement: Maria would support Bronislawa while the latter obtained a degree, then the older sister would reciprocate. During the next five years, Maria worked as a tutor and a governess, falling in love with a man but being heartbroken when the family, distantly related through Maria's father, rejected the notion of marriage.

By 1891, Maria, who would become known as Marie in France, joined her sister and brother-in-law in Paris and enrolled at the Sorbonne. She was introduced to

CURIE RECEIVED TWO NOBEL PRIZES IN HER LIFE, ONE IN PHYSICS AND ONE IN CHEMISTRY

TRAGIC DEATH OF PIERRE CURIE

On a rainy 19 April 1906, Pierre Curie had just finished lunch with a few professional associates in Paris, and was walking to another appointment nearby. When he reached the intersection of the Quai des Grands Augustins with the Rue Dauphine near the Pont Neuf, he attempted to quickly cross one of the most dangerous intersections in the city. Reportedly, two police officers were stationed at the intersection at all times to direct traffic. However, if they were present on this day there was little that could have been done to prevent the tragic accident that occurred. One of the world's foremost physicists stepped into the path of a horse-drawn cart and was immediately struck, apparently falling beneath the wheels of the cart and fracturing his skull. He died swiftly.

When she received the news of her husband's death, Marie was heartbroken, but maintained her composure. Others attributed the cause of the accident, at least in part, to Pierre's carelessness and hurried pace. When his father learned of the tragedy, he responded: "What was he dreaming of this time?" A lab assistant had reportedly observed that Pierre was often inattentive while walking and riding his bicycle, "...thinking of other things."

Pierre Curie contributed to the successes in the early study of radiation, and died an untimely death

Images source: Adobe Stock, Getty, Wiki

Marie and Pierre Curie sit with their eldest daughter, Irène in 1902. A second daughter, Ève, was born in 1904

Marie and Pierre Curie pose for a photographer in their laboratory. The couple became famous for their scientific work

a community of physicists and chemists who were already establishing their own preeminence in these fields. Inspired, Marie worked tirelessly to obtain licensing in physical sciences and mathematics, and assisting in the laboratory of physicist and inventor Gabriel Lippmann, a future Nobel laureate. The long hours took their toll on Marie's health, as she subsisted primarily on tea, bread and butter. In three years, though, she had achieved her immediate goals.

By 1894, Marie had secured a commission from the Society for the Encouragement of National Industry for research on the magnetic properties exhibited by numerous types of steel. Supposedly she needed a laboratory to work in, and was introduced to Pierre Curie by fellow physicist Józef Wierusz-Kowalski. Pierre made room in his personal living space, and a romance developed, but Marie returned to Poland that summer to visit her family, and hoped to gain a respectable teaching position at Kraków University. Gender discrimination again stood in her way, and Pierre persuaded her to return to Paris. The couple married on 26 July 1895, and a pivotal scientific partnership was poised for great achievement.

During this golden age of rapid scientific discovery, Marie searched for a worthy topic for further research, and the production of a thesis. German engineer and physicist Wilhelm Röntgen discovered the presence of x-rays in 1895, and the following year French physicist Henri Becquerel detected emissions of similar rays while researching uranium. News of the discoveries intrigued Marie. The rays did not depend on an external energy source. Apparently, they were produced within the uranium itself.

Using a spectrometer that Pierre and his brother had developed 15 years earlier, Marie determined that the level of activity present was solely dependent on the quantity of uranium being studied, and the activity remained constant regardless of the form of the element. She concluded that the energy was a product of the atomic structure of uranium rather than interaction between molecules, thereby giving rise to the field of atomic physics.

"THE COUPLE MARRIED ON 26 JULY 1895, AND A PIVOTAL SCIENTIFIC PARTNERSHIP WAS POISED FOR GREAT ACHIEVEMENT"

Marie named the newly discovered form of energy 'radioactivity', and began researching other minerals that exhibited similar properties. She found that the mineral pitchblende, now known as uraninite, was ideal for continued research. Pierre discontinued his work on other projects and joined Marie. In the summer of 1898, the husband and wife team discovered the element polonium, which Marie named after her homeland of Poland. Later that year, they discovered a second element and called it radium. Pierre concentrated on the physical properties of radioactivity, while Marie worked to isolate radium in its metallic state.

Meanwhile, the Curies and Becquerel were jointly awarded the Nobel Prize in Physics in December 1903, in recognition of their collective research on 'the radiation phenomena' discovered by Becquerel. Although the initial nomination was intended for Pierre Curie and Becquerel only, Pierre's complaint to the Royal Swedish Academy of Sciences resulted in the addition of Marie to the award as the first woman to receive the Nobel Prize. The Curies also received the prestigious Davy Medal from the Royal Society of London that year.

In 1906, Pierre was killed in an accident on a Paris street. Marie was devastated, but pursued her research with renewed vigour, succeeding her late husband as chair of the physics department at the University of Paris. Four years later, she successfully isolated radium as a pure metal. Even as her personal life became embroiled in scandal during a period of French xenophobia because of her foreign birth, right-wing criticism of her apparent atheistic perspective on religion, and speculation that she was Jewish amid a rising tide of anti-Semitism, her scientific contributions were undeniable. In 1911, it was revealed that she had been involved in an affair with a former student of her husband who was separated from his wife. Nevertheless, in that same year she

Marie Curie pictured here in her laboratory c.1900

Marie gives a lecture on radioactivity in 1925

received the Nobel Prize in Chemistry for the discovery of polonium and radium and the isolation of radium.

As the first person to receive the Nobel Prize twice, and to be so recognised in two separate fields (physics in 1903 and chemistry in 1911), Marie Curie's prestige made a convincing argument for government support of the establishment of the Radium Institute in 1914 at the University of Paris, which continues to this day as a leading research institution in medicine, chemistry and physics. With the outbreak of World War I, Curie worked to establish mobile x-ray units using equipment adapted to automotive chassis. Eventually, 20 of these units were completed. With the help of her daughter, Irène, the mobile units dubbed 'Little Curies' saved many lives thanks to their proximity to the battlefield.

After the war, Marie continued her research in radioactive materials and chemistry. In 1921, she travelled to America to raise funds for the Radium Institute. She was hailed upon arrival in New York City, and attended a luncheon at the home of Mrs Andrew Carnegie and receptions at the Waldorf Astoria hotel and Carnegie Hall. In Washington, DC, President Warren G Harding presented her with a gram of radium and praised her "great attainments in the realms of science and intellect."

Marie gave lectures and became a member of the International Commission on Intellectual Cooperation under the auspices of the League of Nations. She authored a biography of her late husband, and in 1925 returned to her homeland to assist with the establishment of the Radium Institute in Warsaw. She travelled to the US again in 1929, successfully raising funds to equip the new laboratory, which opened in 1932 with her sister, Bronislawa, as its first director. Prior to the development of the particle accelerator in the 1930s, continuing atomic research depended upon the availability of radioactive materials. Marie realised the importance of maintaining adequate stockpiles, and her advocacy facilitated discoveries by Irène and her husband, Frédéric Joliot-Curie.

Years of prolonged exposure to radioactive materials took their toll on Marie's health. Little was known of the effects of radiation exposure – she would carry test tubes of radioactive material in the pockets of her dress, and store them in her desk drawer. She was said to have commented on the soft glow emitted from the tubes, but never realised their potential lethality. During World War I, she had also been exposed to radiation while operating x-ray equipment. As early as 1912 she had been temporarily incapacitated with depression and undergone surgery for a kidney ailment. As a result of radiation exposure, she developed leukaemia and died in Paris at the age of 66 on 4 July 1934. She was buried beside her husband.

Marie Curie remains a towering figure in the fields of physics and chemistry. Her ground-breaking achievements were also empowering for succeeding generations of women, in science and other disciplines.

A DAUGHTER'S CONTRIBUTION

Irène Joliot-Curie, the oldest of two daughters of Pierre and Marie Curie, was an eminent scientist in her own right. Along with her husband, Frédéric Joliot-Curie, she received the 1935 Nobel Prize in Chemistry for her research into the properties of the atom.

The couple's greatest discovery resulted from the exposure of previously stable material to radiation, which in turn caused the material itself to become radioactive. The scientists made the discovery after bombarding a thin strip of aluminium with alpha particles, in this case helium atom nuclei. When the external source of radiation was removed, the aluminium continued to emit radiation because the aluminium atoms had been converted to an isotope of phosphorus. The discovery of artificial radiation served as a catalyst for further research into radiochemistry and the application of isotopes in medical therapies, and largely replaced the costly process of extracting radioactive isotopes from ore. The work of Irène and Frédéric also contributed to the discovery of the process of nuclear fission.

In later years, Irène became the director of the Radium Institute in Paris, and both were leaders in the development of atomic energy in France. She died of leukaemia in 1956 after years of exposure to radiation.

Irène Curie and her husband, Frédéric Joliot-Curie, conducted landmark experiments resulting in the discovery of artificial radiation

While still a young scientist Niels Bohr began to make interesting contributions to the theory of atomic physics

NIELS BOHR

THE ATOM WAS POORLY UNDERSTOOD BEFORE NIELS BOHR CAME UP WITH A MODEL THAT REVEALED THEIR TINY WORLD

When Niels Bohr was born in Copenhagen in 1885 it seemed as if life would be very easy for him. His father was a professor of physiology and his mother came from a wealthy banking family. His father actively encouraged Bohr to study physics at university as he had an interest in it as well as natural talent. There was only one professor of physics at Copenhagen University at the time but Bohr was soon taken under his wing.

Quantum leaps

Bohr proved to be a brilliant student. While still an undergraduate he won a gold medal from the Academy of Sciences for his work. His research started in the practical and experimental world of classical physics but he soon moved into theoretical work. For his PhD Bohr worked on studying how electrons move within metals. While he could explain many of their behaviours it became apparent to Bohr that traditional models of electrons were insufficient, and that a new theory was needed to explain them.

In 1911 Bohr moved to England to work at the universities of Cambridge and Manchester. He collaborated with some of the most eminent atomic physicists of the day. In 1913, while working as a lecturer in Copenhagen, Bohr published a trilogy of papers which revolutionised our understanding of the atom.

Before Bohr there had been several models of atomic structure. Some saw the atom as a lump of positively charged material with negatively charged electrons stuck to it. Ernest Rutherford however had discovered that the nucleus of an atom was very small and the electrons must be outside it somehow. Bohr suggested that the electrons orbited the nucleus of the electron in the same way that planets orbit the Sun. Different orbits have different energy levels and by either taking in energy or giving it off electrons were able to leap between these quantum levels. Many of the most difficult problems to understand in quantum physics and chemistry were solved by this idea.

Nobel and Nazis

In 1922 Bohr was awarded the prestigious Nobel Prize in physics for his creation of the Bohr model of the atom. Still relatively young, Bohr continued to work at the forefront of physics. Quantum physics, the theory of the smallest particles, was still being developed. Bohr became convinced that photons, single bits of light, existed as both waves and particles at the same time.

Bohr collaborated with many of the greatest scientists of his day, including Albert Einstein, and revolutionised atomic theory

Bohr was the director of the Danish Institute of Physics and continued working on atomic theory, such as how atoms can break apart in nuclear fission. With the rise of Nazi power in Germany in 1933 it became dangerous for Jewish scientists to live there. He offered refuge to many by finding them jobs in Denmark and other countries. When Germany invaded Denmark in 1940, Bohr was endangered as his mother was Jewish. Bohr was smuggled to safety in Sweden in 1943, and he persuaded the Swedish government to help other Jewish refugees.

Bohr returned to Denmark after the Second World War and was instrumental in setting up several groups to help European scientists work together – including the famous CERN in Switzerland where the Large Hadron Collider continues to explore the mysteries of the quantum world.

ATOMIC BOMBS

During the Second World War physicists became convinced that it was possible for nuclear fission to power atomic bombs of incredible power. Niels Bohr was taken to America to help with the creation of atomic weapons due to his expertise in nuclear theory. Bohr did offer theoretical advice on the idea but claimed not to have helped much with the final design. Bohr became convinced that nuclear weapons were a danger to the world and worked towards international cooperation.

Bohr's work influenced the creation of nuclear weapons

Images source: Adobe Stock, Wiki

ALBERT EINSTEIN

THIS ICONIC SCIENTIST CHANGED OUR VIEW OF THE WORLD AND HIS NAME HAS BECOME A BYWORD FOR GENIUS, BUT WHO WAS THE MAN BEHIND THE MOUSTACHE?

WRITTEN BY **JODIE TYLEY**

In 1905, Albert Einstein published four papers that revolutionised our understanding of the universe. He won a Nobel Prize for his contribution and came up with the most famous scientific equation: E=mc2 - as recognisable as his distinctive face. At the time of these breakthroughs, however - a year that would become known as his 'annus mirabilis', or 'miraculous year' - he was a dark-haired, doe-eyed 26-year-old. Handsome and known for being a bit of a ladies' man, he didn't even have a PhD. In fact, he was working in a Swiss patent office, a role significantly less prestigious than his desired doctorate.

However, the position afforded him time to theorise on the properties of light. Einstein worked best as an independent thinker, which is one of the reasons for his troubledexperience of education. Likening his teachers to "drill sergeants," he earned a reputation as a mischief-maker. When his father asked why, the teacher said he "sits at the back and smiles." There was no way his parents could have fathomed what a genius he would become. He was slow to develop, beginning to speak some time after the age of two. "My parents were so worried," he later recalled, "that they consulted a doctor."

Even when he did start to communicate, he would whisper the words to himself first, perfecting the sentence until it made sense to say out loud. The family maid called him 'der Depperte' - the dopey one - and they thought he'd never be a model student. Einstein believed this allowed him to ponder things that others took for granted. When his father handed him a magnetic compass to relieve his boredom, the five-year-old was fascinated by the invisible forces acting upon the needle, turning it this way and that. Something so intriguing had never been discussed in school, and Einstein quickly realised he would have to work things out for himself. "I have no special talents," he later declared, "I am only passionately curious."

As a youngster, he enjoyed puzzles and building houses of cards with the help of his adoring little sister, Maja. Before she was born, his mother said he would soon have a wonderful toy to play with. "Where are the wheels?" he famously exclaimed when he was presented with a chubby newborn. The pair would become incredibly close, though, despite his childhood tantrums when he would hurl objects at her head. "It takes a sound skull to be the sister of an intellectual," she later joked of their relationship.

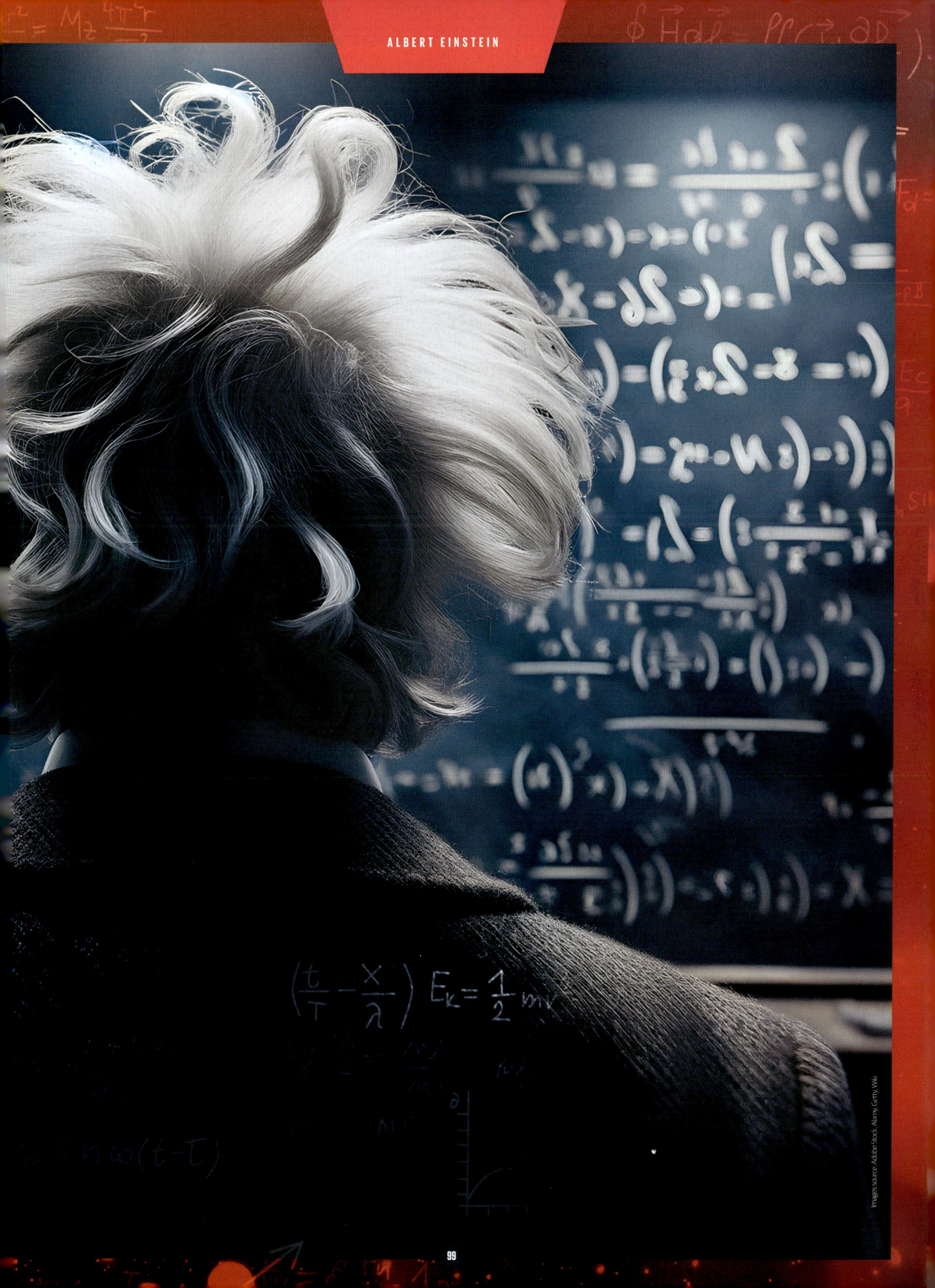

Images source: Adobe Stock, Alamy, Getty, Wiki

The turning point for Einstein came during his teenage years, while a medical student lodged with his family. Max Talmey introduced him to algebra and gave him books on geometry and the natural sciences. One particular volume by Aaron Bernstein described a current of electricity racing down a telegraph wire and asked the reader to imagine running alongside it, which led Einstein to ponder the nature of light. If you could catch up to a beam of light, he thought, it would appear frozen, but no one had observed this before.

Famous for thought experiments such as this, he preferred to deliberate in pictures rather than words. In 1904, a 25-year-old Einstein would walk the streets of Bern, Switzerland, with his baby son, Hans-Albert, in a stroller. It was nine years since he had read Bernstein's book, but the puzzle had always stayed with him. Brow furrowed and determination etched into his features, he would pause to take out the notepad that lay next to the tiny infant and scribble down a series of mathematical symbols. The *New York Times* would later sum this momentous event up best, commenting that: "Out of those symbols came the most explosive ideas in the age-old strivings of man to fathom the mystery of the universe."

At the end of the 19th century, light was assumed to be a wave travelling through a mysterious thing called aether. Einstein removed aether from the equation entirely with his general theory of relativity, creating a fundamental link between space and time. He explained that time passes at different rates depending on how fast an object is moving; the faster it travels, the slower time progresses. The equation E=mc2 represents the relationship between mass (m) and energy (E). Essentially, Einstein found that when an object approached the speed of light (c), the mass of the object increased. Or, as he put it most simply: "When you are courting a nice girl, an hour seems like a second. When you sit on a red-hot cinder, a second seems like an hour. That's relativity."

When considering how he happened to be the one to come up with such a groundbreaking theory, he explained that he owed it to those early years that caused his parents such concern. "The ordinary adult never bothers his head about the problems of space and time. These are things he has thought of as a child. But I developed so slowly that I began to wonder about space and time only when I was already grown up. Consequently,

Einstein with his second wife Elsa outside the State, War and Navy Building in Washington, DC

Einstein, aged 42, giving a lecture in Vienna in 1921

I probed more deeply into the problem than an ordinary child would have."

When the news of Einstein's findings broke, the media turned him into a global sensation. It was the new theory that everyone was talking about and no one understood, and arrived in a period of great social change. World War I had ended the year before, there were new technologies being developed and the roles of the sexes were being rebalanced. Professionally, things had never been better for the scientist, but behind closed doors, Einstein's personal life was crumbling.

His 11-year marriage to Mileva Marić was falling apart and he issued her an ultimatum - if they were to remain together for the children, she would have to agree to a list of conditions. From making sure his laundry was kept in good order to leaving his study immediately if he requested it, they were pragmatic and cold demands. Months later, Marić returned to Zurich with their two sons. Einstein's first born, Hans-Albert would grow up and reflect that "probably the only project he ever gave up on was me." His father did see that they were looked after financially, giving his Nobel Prize winnings to the family.

Following the divorce, Einstein married his cousin and long-time mistress Elsa Löwenthal, but it would seem his only true love was science. Between the mid-1920s and his emigration to the USA in 1933, there were half a dozen women in his life. The fact that Einstein's move to the USA coincided with Hitler's rise to power is no coincidence. A hateful anti-Semitic campaign was set up by the Nazis to discredit the Jewish scientist and his theories. They painted him as a fraud, suggesting he plagiarised his work, and these ill-grounded accusations have plagued accounts of Einstein's life ever since.

But just as Nazi Germany was suspicious of Einstein, he was deeply wary of them. Believing they were developing an atomic bomb, he wrote to the American President Roosevelt to warn of a growing nuclear threat. He encouraged the US government to research nuclear chain reactions using uranium in response to German advances in the field. But Einstein was also a life-long pacifist and opposed the war. It was the reason he left Germany aged 16. The law required every German male to serve in the military, so his only option was to leave before his 17th birthday. After a careful escape, he joined his parents in Italy and renounced his German citizenship.

With the significant weight of his reputation, it can be said that this letter to Roosevelt was the catalyst for the USA's development of the atomic bomb. The Manhattan Project was fatefully begun, but Einstein himself never worked directly on it. Nevertheless, his famous equation inadvertently provided the starting point for its research, and he would be doomed to forever explain his unwitting role in this pivotal moment in history.

Adamant that all he had done was write a letter, Einstein came to regret even sealing the envelope. In an interview with *Newsweek* magazine, he said: "Had I known that the Germans would not succeed in developing an atomic bomb, I would have done nothing." When the first atomic bomb was dropped on Hiroshima, Japan (killing up to 140,000 people), this action and its aftermath led to him undertaking anti-nuclear campaigns and lectures for the rest of his life. When asked what weapons World War III would be fought with, he famously replied that he did not know, commenting portentiously however that "World War IV will be fought with sticks and stones."

Einstein's later years saw him pioneer numerous key theories including wormholes, multi-dimensional models and the possibility of time travel, as well as discovering his unified field theory - an all-embracing theory that would unify the forces of the universe and physics into one framework. It was never completed, however, as Einstein died of an aortic aneurysm in 1955. He had refused surgery, saying: "I have done my share, it is time to go. I will do it elegantly." Sadly, as his nurse didn't speak German, we'll never know what his last words were. They are lost to space and time, which we understand a lot better thanks to him.

EXPANSION OF THE UNIVERSE

A TIMELINE OF THE UNIVERSE FROM ITS CREATION, ABOUT 13.8 BILLION YEARS AGO, TO TODAY

Today

Modern galaxies

1 billion years

Early galaxies

First stars appear

300 million years

380 thousand years

Big Bang

0

According to recent work on the theory, very shortly after the initial singularity, the universe went through a period of hyper-quick expansion, unconstrained by the speed of light, before settling down to a slower but continuing growth

IT BEGAN WITH A BANG

HOW A GROUP OF UNORTHODOX SCIENTISTS, LED BY A CATHOLIC PRIEST, UPTURNED SCIENTIFIC BELIEFS ABOUT THE START OF THE UNIVERSE

WRITTEN BY **EDOARDO ALBERT**

Einstein later changed his mind about the work, calling it the "the most beautiful and satisfactory explanation of creation to which I have ever listened"

In March 1949, English astronomer Fred Hoyle gave a talk on BBC radio about the competing views of the origin of the universe. There was one which proposed that "all the matter in the universe was created in one big bang at a particular time in the remote past". Hoyle himself thought this particular theory was nonsense but, unwittingly, in that radio talk he gave the name to what would become the dominant explanation of how the universe began.

The man who first proposed the theory was Georges Lemaître (17 July 1894 - 20 June 1966). Physics was not Lemaître's main job: he was a Catholic priest and a Jesuit. As such, he saw no conflict between science and faith. Not all of his fellow scientists agreed. Indeed, much of the opposition to his idea of the universe having started as a "primeval atom" came from scientists who viewed the idea as smuggling creationism into physics. They much preferred Fred Hoyle's idea of a 'steady-state universe'.

Both Lemaître's theory of the Big Bang and Hoyle's steady-state universe set out to explain one of the most extraordinary findings of early 20th-century science: we appeared to be Billy No-Mates. All the galaxies astronomers could observe appeared to be running away from our own Milky Way and, what's more, the further away they were, the faster they were running. It's as if everything in the universe was trying to get away from us.

Hoyle, and most other scientists at the time, subscribed to a belief that the universe had no beginning: it had always existed and it would continue to always exist. Even Einstein initially thought Lemaître's theory was nonsense. To explain the apparent expansion revealed in the early 20th century, Hoyle postulated that matter was spontaneously created, literally springing into being out of nothing at a steady rate. This creation of matter allowed for the universe to remain in a steady state despite its expansion (otherwise, with an infinitely old universe, we would find ourselves alone in an absolute void).

Lemaître, on the other hand, proposed that at some point in the past the entire universe had been concentrated together in one tiny point from which it had expanded, and was continuing to expand after, as Hoyle put it, a Big Bang (indeed, the Biggest Bang there could ever be).

The discovery of cosmic microwave background radiation in 1964 by Arno Penzias and Robert Wilson provided strong evidence for Lemaître's Big Bang theory. This is a bath of microwave radiation that pervades all the universe in a sort of afterglow of its violent beginning. It turned out that Lemaître was right (although Hoyle never accepted this).

Lemaître did not mix his religion and his science but it turns out that, in the beginning, there was nothing. And then, there was light.

Images source: Adobe Stock, Getty

UP, UP AND AWAY

FROM KITES TO JET ENGINES, THE HUMAN QUEST FOR FLIGHT HAS BEEN THOUSANDS OF YEARS IN THE MAKING

ANCIENT AVIATION

Invented in China during the Warring States period, kites were mainly used for military purposes, such as communication, measuring distances and calculating wind readings. They eventually spread throughout Asia and remain popular to this day.

C. 400 BCE

FULL STEAM AHEAD

While the inventor Daedalus and his high-flying son Icarus were mythical, the real-life ancient Greek engineer Hero of Alexandria developed the aeolipile, a primitive steam turbine that used similar principles to today's jet propulsion.

50 CE

FLIGHT AND FIGHT

The first ever aerial bombing took place against Libya by Italian lieutenant and pilot Giulio Gavotti during the Italo-Turkish War, marking the beginning of aerial warfare.

1911

WORLD'S FIRST COMMERCIAL AIRLINE

3 MONTHS

The St. Petersburg-Tampa Airboat line did not last long

Number of people who gathered to watch the airline's inaugural flight

3,000

A passenger ticket cost $5 each way

$5

1914

ADVANCING AVIATION

During World War I, aerial warfare became more important than ever. Initially unarmed, aeroplanes started to carry machine guns and even explosives, while new types of aircrafts such as night bombers emerged.

At the beginning of the war, aeroplanes were mainly used for reconnaissance purposes

1914 - 1918

THE HINDENBURG DISASTER

The German zeppelin exploded in Lakehurst, New Jersey, killing 35 people on board and one person on the ground. With photographs, news footage and eyewitness reports of the tragedy widely shared, the dreams of airship travel quickly ended.

1937

THE BATTLE OF BRITAIN

1,000

Approximate number of German aircraft that headed to London on the first day of the Blitz

1,547

Number of aircraft lost by the Allies

1,887

Number of aircraft lost by the Germans

1940 - 1941

THE FIRST JET FIGHTER

Nazi Germany hoped that the Messerschmitt Me 262, one of the most advanced aircraft of World War II, would turn the tide of the war in their favour – but it was introduced too late and insufficient numbers were made to make a difference.

1941

SPEED AWAY

US Air Force Captain Chuck Yeager becomes the first human to fly faster than the speed of sound and therefore break the sound barrier. He went on to break several more speed records during his career.

1947

Da Vinci produced a sketch for an aerial screw, the predecessor to the helicopter

FLIGHT OF FANCY

Leonardo Da Vinci was fascinated with aviation. He sketched designs for an ornithopter, a flying machine where the pilot would lie in the prone position and use a crank to control a rod and pulley system to move the wings.

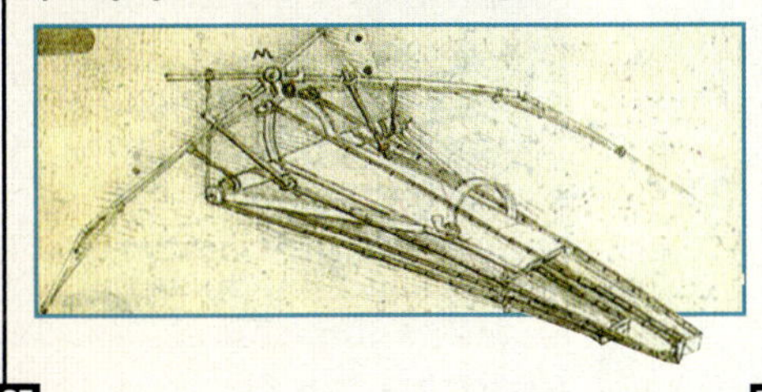

1485

UP IN THE AIR

French chemistry teacher, Jean-Francois Pilatre de Rozier completed the first manned hot air balloon flight, before a crowd of dignitaries in Annonay, France, in the aircraft developed by the Montgolfier brothers.

1783

GLIDING ACROSS THE SKY

Sir George Cayley spent over 50 years on aeronautical experiments, notably on his gliders, determined to find a way for humans to fly and was the first person to conceptualise the modern airplane with a fuselage, wings and a tail.

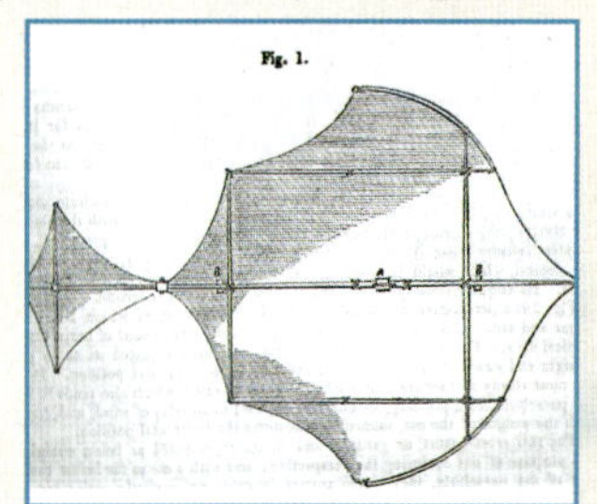

1799 – 1850

PROPELLED INTO ACTION

The Breguet-Richet Gyroplane I lifted off the ground for the first time in September. Built by brothers Louis and Jacques Breguet alongside Professor Charles Richet, it was considered the first manned flight of a helicopter.

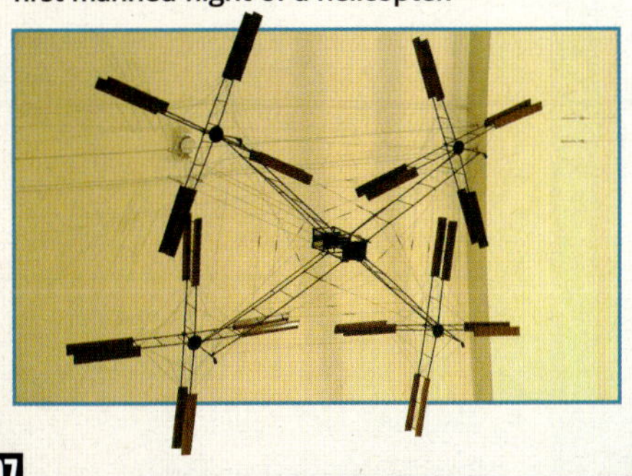

1907

MAKING HISTORY

After working towards it for several years, the Wright brothers completed the first successful, controlled and sustained flight of their biplane, the Wright Flyer, on Kitty Hawk Beach, in North Carolina.

1903

THE SIEGE OF PARIS

66 Number of hot air balloons that flew out of Paris during the siege of city by Prussian forces

2.5-3 MILLION Approximate number of letters carried by the balloons

110 Number of passengers who also boarded the balloons

1870

REACHING NEW HEIGHTS

When the Soviet Union launched Sputnik 1, the first man-made satellite to orbit the Earth, it proved that the sky was no longer the limit for aviation.

Sputnik's launch sparked the Space Race between the United States and the Soviet Union, leading to the Moon landings

1957

CONCORDE TAKES OFF

The supersonic aircraft had a take-off speed of 250MPH

It had a cruising speed of 1350MPH more than twice the speed of sound

3.5 HOURS The time it took for Concorde to fly from London to New York

1976 - 2003

NEW INNOVATIONS

The Solar Impulse 2 became the first solar-powered aircraft to successfully circumnavigate the globe in just over 16 months, demonstrating the viability of renewable energy for aviation.

In 2017, Google co-founder Larry Page funded a prototype for an all-electric flying car, named the Kitty Hawk

2015

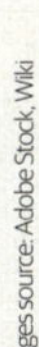

PIONEERS OF COMPUTING

THE MEN AND WOMEN WHOSE INVENTION AND INNOVATION HELPED TO SHAPE THE MODERN WORLD

TIM BERNERS-LEE

ENGLISH, 1955 – PRESENT

Born to parents already working in computing, Berners-Lee took to the industry quickly, writing software after graduating university in 1976 before joining CERN as a software engineering consultant in 1980. While there he developed hypertext for containing links between files he was working on. Returning to CERN in 1984 he developed the concept of a global hypertext document system using HyperText Markup Language (HTML) and a Uniform Resource Locator (URL), the building blocks of the web.

ALAN TURING

ENGLISH, 1912 – 1954

A number of important milestones in computing can be attributed to Alan Turing. While war records had been kept secret for some time, he is now best remembered for his contribution to the Bletchley Park code-breaking efforts during WWII.

His mathematical insight helped to design the cipher machines needed to break the German encryption of classified intelligence. Prior to this, Turing had conceived of the Turing machine, a theoretical computer, which he later developed into the Automatic Computing Engine from 1945. The design was deemed too complex to attempt, but lit the way for future computing innovations.

Turing also worked on the concept of artificial intelligence, developing the Turing Test to judge a machine's abilities.

KONRAD ZUSE

GERMAN, 1910 – 1995

If not for having developed his ideas in Germany during WWII, Zuse might well have been a much more famous name in the world today. He developed the Z3 in 1941, the first functional program-controlled computer. However, it wasn't widely adopted and was destroyed in a bombing raid in 1943. Zuse founded his own computer business and built the Z4, beaten to market only by the UNIVAC I in 1951 as a commercial computer.

DOUGLAS ENGELBART

AMERICAN, 1925 – 2013

Two of the most important innovations for how we use computers today are the graphical user interface (GUI) and mouse used to interact with it, and we have Engelbart, in part, to thank for both. In 1963 he received partial funding for his research into computer interfaces and in 1968 he presented a collaborative computer system using mouse and keyboard. The presentation of the work of the Stanford Research Institute (SRI) also included editing shared documents. It became known as 'the mother of all demos'.

CHARLES BABBAGE

ENGLISH, 1791 – 1871

A mathematical visionary, Babbage was self-taught in algebra and influenced by the mathematics being discussed on the Continent. Following university he became a fellow of the Royal Society in 1816 and helped found the Astronomical Society in 1820, from which his interest in mechanical calculation began. He invented the Difference Engine for making mathematical tables in 1821 and then the Analytical Engine for more complex mathematical equations, storing up to 1,000 50-digit numbers. Technology of the time couldn't build a lasting Difference Engine and the Analytical Engine wasn't completed, but they set the template for computer technology to come.

Along with his computing inventions, Babbage also created the locomotive cowcatcher.

ADELE GOLDBERG

AMERICAN, 1945 – PRESENT

Helping to further advance the way we interact with computers, Goldberg's work with Xerox PARC (Palo Alto Research Center) resulted in Smalltalk-80, which led to a graphical user interface (GUI) that included many things we take for granted today. This included windows, icons, menus and a pointer for selecting them. Smalltalk-80 was demonstrated to Steve Jobs in 1979 and is believed to have inspired the user interface later Apple computers employed.

GRACE HOPPER

AMERICAN, 1906 – 1992

You might not imagine someone could be made a rear admiral in the US Navy as a mathematician, but that's exactly what Grace Hopper achieved. She joined the Navy as a reservist while teaching at Vassar College in 1943 and was called up for duty, working on the Mark I from 1944, the first large-scale automatic calculator. From here she began her trailblazing journey in computer languages. One of her key contributions was in creating the first compiler that translated mathematical code into machine binary, the first steps to a universal language for writing programs.

Hopper was originally rejected by the War Office as a reservist due to being too short.

STEVE WOZNIAK

AMERICAN, 1950 – PRESENT

Known affectionately around the world by tech aficionados as 'Woz', Steve Wozniak was the co-founder of Apple Computer with Steve Jobs in 1976. The pair had been members of a homebrew computer club and having had his design for a new microcomputer rejected by his employer Hewlett-Packard, Wozniak teamed up with Jobs to sell it themselves as a kit computer. It was so successful that they subsequently designed the Apple II, released in 1977, which became one of the very first successful personal computers for the mass market. Computing was now accessible to all like never before.

ADA LOVELACE

ENGLISH, 1815 – 1852

"The Analytical Engine weaves algebraic patterns just as the Jacquard loom weaves flowers and leaves," wrote Lovelace in 1843. She met Babbage in 1833 and came to translate a paper on his Analytical Engine by Italian engineer Luigi Menabrea in which she praised his creation. It was in the notes to this translation that Lovelace wrote the first ever example of a program for such a device, making it the first computer program ever written. But, as her quote above suggests, she saw the broader creative possibilities of such technology as well as how it would be limited by our own knowledge.

Since 2009, the second Tuesday in October is Ada Lovelace Day, celebrating women in STEM.

JOHN VINCENT ATANASOFF

AMERICAN, 1903 – 1995

In 1973, Atanasoff saw his Atanasoff-Berry Computer, developed between 1937 and 1942 with his graduate student Clifford Berry, successfully recognised in court as the first electronic digital computer. Atanasoff's work had begun when he found current analog machines for calculation too limited for his differential equation work. The device he created could hold data through capacitors and used logic circuits for performing basic functions.

Images source: Adobe Stock, Getty Images

The storage capacity of the Analytical Engine was larger than any computer built before 1960.

1821 DIFFERENCE ENGINE

Frustrated by the inaccuracy of printed calculation tables used in mathematics, statistics and more, Charles Babbage conceives of a mechanical calculating engine that would be infallible. This develops into the bigger concept of a general purpose Analytical Engine in 1834. Neither is actually built in his time.

1936 THE TURING MACHINE

Alan Turing lays out his idea for a computing device capable of being programmed for any mathematical problem. The intention is to be able to tackle mathematics' most complex or undecidable problems. This hypothetical machine inspires the computing designs that follow.

THE ANTIKYTHERA MECHANISM c.100 BCE

Designed to display details of astronomical phenomena through a series of gears, this ancient Greek device is thought to be the earliest known example of a basic 'computer'.

THE FIRST PROGRAM 1843

Translating a paper on Charles Babbage's work by Luigi Menabrea, Ada Lovelace includes extensive notes on how to program the Analytical Engine, writing the first computer program.

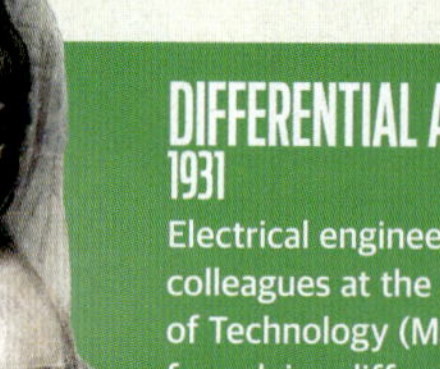

DIFFERENTIAL ANALYZER 1931

Electrical engineer Vannevar Bush and colleagues at the Massachusetts Institute of Technology (MIT) create a machine for solving differential equations called the Differential Analyzer.

1821 1936 1943

A WORD IS BORN 1613

Author and poet Richard Braithwaite uses the word computer in his book *The Yong Mans Gleanings*. This is the first recorded use of the word, describing a mathematician.

SPEED AND EFFICIENCY 1890

With the previous census having taken seven years to collate, the US government turns to Herman Hollerith's electronic tabulator. He completes the tabulation before the end of the year, saving taxpayers $5 million.

LOST TO WAR 1941

The first automatic computer, running on telephone relays and controlled by programs, is revealed in Berlin by Konrad Zuse. It is destroyed during Allied bombing in 1943.

1943 THE COLOSSUS

Considered to be the first semi-programmable digital computer in the world, the Mark I Colossus is delivered to Bletchley Park to assist in cracking the German Lorenz cipher. Unlike Enigma, Lorenz was used to encrypt much longer messages discussing longer-term plans and intentions.

By the end of WWII the Colossus machines had helped to decrypt 63 million characters of high-level German communications.

1969 NETWORKED COMMUNICATION

The Advanced Research Projects Agency Network (ARPANET) is developed by the US Department of Defence for cross-departmental collaboration and communication. Its cross-national network of computers is a precursor for the worldwide internet to come.

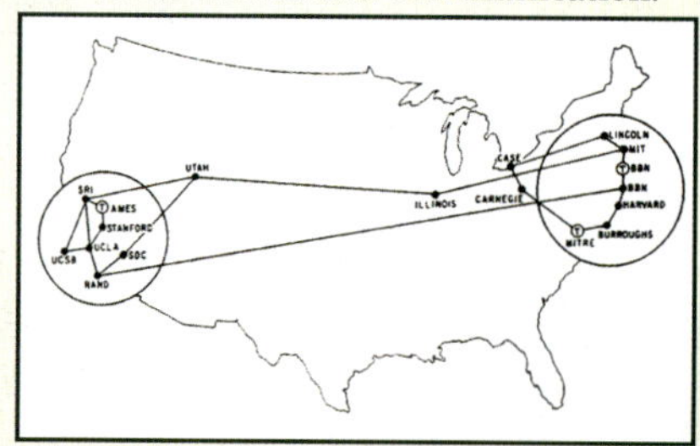

1990 THE INTERNET

While working at the European Organization for Nuclear Research (CERN), British scientist Tim Berners-Lee develops the concept for an automated information-sharing system. Working with systems engineer Robert Cailliau, they lay out the basis for the World Wide Web, the beginning of the internet.

The original intent of Berners-Lee's work was to facilitate communication between scientific laboratories.

BUSINESS MACHINE 1954

The new IBM 650 Magnetic Drum Calculator is released for accounting and computation, becoming the most popular computer of the 1950s, selling around 2,000 units.

NEW INTERFACE 1968

American inventor Douglas Engelbart presents his prototype for a modern computer, featuring a mouse and graphical user interface, at the Fall Joint Computer Conference in San Francisco.

MICROSOFT BEGINS 1975

Childhood friends Paul Allen and Bill Gates begin translating BASIC for the kit computer Altair, and start their own company, Microsoft.

1969 1977 1985 1990

NEW LANGUAGE 1959

The COBOL (Common Business-Oriented Language) is developed by a team including Rear Admiral Grace Hopper, based on her own design, to create a common programming language for computers.

GAMING CONSOLE 1972

Ralph Baer releases the Magnavox Odyssey, the first home gaming console. The machine uses cartridges so that games can be swapped out and dial-based control devices.

CHECKMATE 1997

IBM's Deep Blue computer defeats Russian grandmaster Garry Kasparov at chess in 37 moves. This is the first computer win against a world champion under tournament rules.

1977 APPLE COMPUTERS

The Apple II is presented by Steve Jobs and Steve Wozniak at the West Coast Computer Faire, beginning a boom in personal computer (PC) sales. They follow it up in 1983 with the Apple Lisa, one of the first PCs to have a graphical user interface (GUI), including features like drop-down menus and icons.

Despite the influential design innovations of the Apple Lisa, it was not a commercial success.

1985 OPENING WINDOWS

The first GUI from Microsoft is built on top of MS-DOS, the computer interface that Microsoft had pioneered for personal computers. It makes specific use of mouse controls, which is still not a universal feature of PCs at the time, and features a game called *Reversi*.

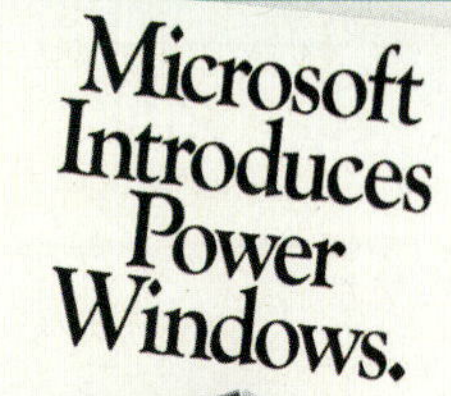

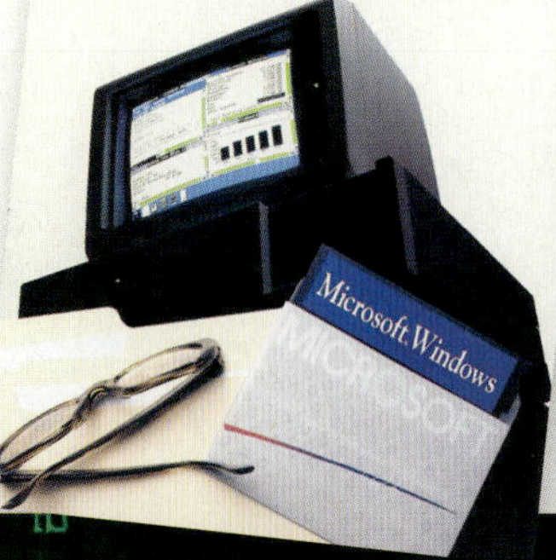

RISE OF THE SPACE AGE

SHADOWED BY FEAR OF WAR, THE EARLY DAYS OF EXPLORATION SHOWED HUMANITY AT ITS MOST BOLD

WRITTEN BY **BEN EVANS**

Over 60 years ago, the world gazed at the sky and listened through shortwave radio receivers with fascination and fear. For millennia, humans had clung to the Earth's surface, only recently having mastered the long-held dream of flight and with scant awareness of what lay beyond the thin veil of the atmosphere. But, on 4 October 1957, our sense of place in the cosmos changed forever. Over three weeks, a steady 'beep-beep' transmission from Sputnik 1 - the first artificial satellite - heralded the dawn of the Space Age. Yet the euphoria of conquering space was met by harsh Cold War reality, as Russia and America sought to deliver weapons of enormous destruction across intercontinental distances.

For something that changed the world, Sputnik 1 was an unremarkable icon. It was a polished metal sphere, 23-inches across, with four antennae to broadcast radio pulses at 20.005 MHz and 40.002 MHz, easily audible to amateur radio listeners. Orbiting at 65-degrees of inclination, its flight path carried it over virtually the entire inhabited Earth every 96.2 minutes. Its signal vanished when its batteries died, and the 184-pound satellite burned up in the atmosphere in January 1958.

Thus began the Space Race between the capitalist United States and the communist Soviet Union to attain mastery over the heavens. Following the World War II, both nations used captured German scientists and rockets (including the infamous V-2) to further their ambitions of building intercontinental ballistic missiles to establish technological and ideological supremacy over the other. Juxtaposed against this bellicose stance was the 1957-1958 International Geophysical Year, a concerted 18-month campaign of Earth science research. In the summer of 1955, the United States and the Soviet Union pledged to launch a satellite during the IGY.

Politically, Sputnik 1 was a great shock, and demolished Western perceptions of Russia as a backward nation made up of potato farmers and inept politicians. Science-fiction writer Arthur C. Clarke reflected that on 4 October 1957, the United States became a second-rate world power, while economist Bernard Baruch praised the Soviet "imagination to hitch its wagon to the stars" and stressed that American paranoia was well founded. During his 1960 presidential campaign, John F Kennedy played into this palpable sense of national dread by claiming that Soviet hegemony in space could someday afford them control of the Earth.

After the 'Sputnik Crisis', political figures increasingly spoke of a 'gap' in missile-building technology, with the United States falling behind the Soviet Union. Indeed, the Soviets created the world's first intercontinental ballistic missile - the R-7 - and test flew it across a distance of 3,700 miles, before using a modified version to launch Sputnik 1. Remarkably, the same basic rocket is still used to launch satellites and humans today. The missile gap was promulgated by the Gaither Report in November 1957, which recommended a significant strengthening of US military might. Its figures were exaggerated, but the fiction of a missile gap galvanised America into forming NASA in October 1958, and accelerated the development of rockets to send men into space.

America's ascendancy into space began with disappointment. In December 1957, a Vanguard rocket exploded on the launch pad, triggering a media frenzy. Journalists mocked it as 'Kaputnik', while Soviet delegates to the United Nations tauntingly wondered if the United States needed their aid as an "undeveloped nation". Finally, on 31 January 1958, Explorer 1 became America's first successful satellite. Six weeks later, it was followed by Vanguard 1, disparagingly nicknamed "the grapefruit" by Soviet Premier Nikita Khrushchev. However, the smallness of these early satellites actually belied their rather advanced scientific capabilities. Explorer 1 discovered the Earth's Van Allen radiation belts, while Vanguard 1 remains the oldest man-made object still in orbit today.

The benefits of satellites for a range

To date, 12 people have left their footprints on the Moon

THE GREAT SPACE RACE

FROM TINY SATELLITES TO BOOTS ON THE MOON, HUMANITY TOOK GREAT STRIDES IN A SINGLE DECADE

UNITED STATES	1955	SOVIET UNION
		4 October 1957 Sputnik 1, world's first artificial satellite.
		3 November 1957 Sputnik 2, carried first living creature into orbit.
31 January 1958 Explorer 1, America's first satellite.		
		12 September 1959 Luna 2, first mission to crash-land on the Moon.
	1960	**4 October 1959** Luna 3, first images of the far side of the Moon.
5 May 1961 Alan Shepard, America's first man in space.		**12 April 1961** Yuri Gagarin, first man in space.
20 February 1962 John Glenn, America's first man to orbit the Earth.		
		16 June 1963 Valentina Tereshkova, first woman in space.
3 June 1965 Ed White, America's first spacewalk.		
15 December 1965 Gemini 7 and 6, first rendezvous in space.		**18 March 1965** Alexey Leonov, first spacewalk.
16 March 1966 Gemini 8, first docking in space.		
24 December 1968 Apollo 8, first mission to orbit the Moon.		
21 July 1969 Apollo 11, first piloted landing on the Moon.	1970	

Sputnik's radio signal was easily detecable, even by amateur equipment

of applications - from communications to reconnaissance and navigation to scientific research - had long been recognised, and in December 1958, the first test of a relay was used to broadcast Christmas greetings from US President Dwight D Eisenhower. Two years later, Echo 1 became the world's first passive communications satellite, followed by Telstar, which transmitted television pictures, telephone calls and telegraph images, as well as a live transatlantic feed between the United States and Belgium.

It was Arthur C Clarke who first widely disseminated the idea of putting satellites into 'geostationary' orbit, more than 22,000 miles above the Earth, matching the planet's rotation for worldwide communications. Syncom 3 was the first to reach this high orbit, relaying images from the 1964 Summer Olympics in Tokyo. This laid the foundation for hundreds more communications satellites, which continue to deliver telephone, television, radio and internet services.

Of course, the Cold War inspired less peaceful activities, too, and planning for reconnaissance satellites was set in motion early in the Space Age. However, it was only after the infamous shootdown of Gary Powers' U-2 reconnaissance aircraft in May 1960 that the need for military eyes in space became commonplace. In August of that year, Discoverer 13 became the first satellite to return an object safely to Earth, in the form of a classified film canister. Less than two weeks later, the Soviets brought their Korabl-Sputnik 2 spacecraft, carrying the dogs Belka and Strelka, safely back home. It was the first time that living creatures had been launched into orbit and returned alive.

Sending living creatures, and eventually humans, into space was an important driving force. In November 1957, the Soviets launched Sputnik 2, carrying a dog, Laika. Several animals had already flown above the 62-mile-high 'Kármán line' - the internationally recognised boundary for the edge of space - but three-year-old Laika was first to actually achieve orbit. Following a stressful launch, Laika died within hours when the cabin overheated. Her legacy is that she unmasked some of the unknowns about the survivability of launch, orbital acceleration and the effects of weightlessness. Laika laid the groundwork for the 108-minute orbital flight of Yuri Gagarin, the first man in space, on 12 April 1961.

If Sputnik 1 shocked the world, then Gagarin's mission shocked it again, particularly as it occurred only months into the administration of President John F Kennedy. Matters worsened when CIA attempts to overthrow Fidel Castro failed, leaving Kennedy humiliated, and in need of a means to re-establish his nation's prestige. Although Alan Shepard became America's first man in space on 5 May 1961, his Redstone booster was only capable of a 15-minute suborbital flight. Not until the following year did John Glenn - riding the larger, more powerful Atlas rocket - actually achieve orbit.

Despite such limited spaceflight experience, Kennedy told a joint session of Congress that he intended to direct the United States to land a man on the Moon before the end of the decade. It was a challenging gamble, since lunar exploration had been pioneered by the Soviets. In January 1959, Luna 1 became the first man-made object to reach the Moon, measuring the solar wind, and eventually entering heliocentric orbit. Before the year ended, its follow-up Luna 2 had been intentionally crashed into the surface, and Luna 3 returned

The US did not have a flying start, their Vanguard rocket exploded on the launch pad

It's argued the US eventually overtook the USSR in the Space Race

the first photographs of the Moon's far side, never before seen by human eyes. On 3 February 1966, a Soviet spacecraft, Luna 9, performed the first soft landing on another celestial body.

Russia also held the advantage in human space exploration, flying cosmonauts into orbit for several days, sending the first woman into space, launching the first multi-person spacecraft and executing the world's first spacewalk. However, the pendulum shifted in the mid-1960s, and America took the lead, flying longer missions, performing spacewalks and docking with other spacecraft. Its investment in Kennedy's goal peaked at five per cent of the federal budget. Meanwhile, the Soviets suffered the premature death of their chief rocket designer, Sergei Korolev, and the advantage slipped from their fingers. Yet the dangers of space exploration were ever-present. America lost three Apollo astronauts in a launch pad fire in January 1967 and, just three months later, a Russian cosmonaut plunged to his death when the parachutes on his descending Soyuz spacecraft tragically failed to open.

"THE PENDULUM SHIFTED IN THE MID-1960S, AND TRULY AMERICA TOOK THE LEAD"

In spite of the emphasis on reaching the Moon, both nations also turned their attention further afield, with the United States completing the first flyby of Mars with Mariner 4 in July 1965. The spacecraft's photographs revealed a hostile world, with no evidence of wind or water erosion, and a virtual absence of magnetic field. Soviet missions to the Red Planet were more troubled: three exploded during launch, and another was lost during its outward journey. Mariner 2 flew past Venus in December 1962, while Russia's Venera 3 was first to crash-land on the planet's surface in March 1966. A year later, Venera 4 became the first spacecraft to take direct measurements from another planet's atmosphere, revealing carbon dioxide as Venus' main atmospheric constituent.

The race to the Moon continued unabated. In November 1967, America test-flew its Saturn V lunar rocket for the first time, and the following September, Russia launched the Zond 5 spacecraft around the Moon, carrying a payload which included mealworms, wine flies, plants and a pair of tortoises. They became the first living creatures to venture into deep space, visit our closest celestial neighbour and return safely to Earth.

As the end of the decade approached and the final lap of the Space Race began, CIA intelligence hinted a large Soviet rocket, the N-1, was undergoing final preparations to send a pair of cosmonauts around the Moon. Reconnaissance satellite imagery showed the rocket on its launch pad and, in August 1968, America hurriedly moved to upgrade Apollo 8 from an Earth-orbital flight to a lunar voyage. In just four months, the mission rose from the drawing-board to reality, and astronauts Frank Borman, Jim Lovell and Bill Anders became the first men from Earth to settle into orbit around the Moon.

The N-1, meanwhile, suffered two catastrophic failures in February and July 1969, eliminating the last remaining Soviet hope of somehow getting cosmonauts onto the lunar surface before Neil Armstrong and Buzz Aldrin. Another unmanned spacecraft, Luna 15, sought to tame the impending American triumph by bringing some lunar soil back to Earth, but it ignominiously crashed into the Moon's surface a few hours after Armstrong and Aldrin landed at the Sea of Tranquillity.

With the space race won, political attitudes changed. The Soviets refocused their attention on building long-term space stations in Earth orbit, while America developed the Space Shuttle as a more cost-effective means of reaching space.

Eventually, the two former foes united their efforts in today's International Space Station. And aboard that station on the 60th anniversary of Sputnik 1's success, astronaut Joe Acaba was filled with wonder for the past and excited hope for the future. "Amazing to be on Space Station and reflect on how far we've come," he posted online. "What will the next 60 years bring us?"

SPACE AGE TIMELINE

THE US AND RUSSIA PRESSURED EACH OTHER INTO GREAT ADVANCES

1950-1959

4 October 1957
Sputnik 1, the world's first artificial satellite, spent three months in space and travelled 43 million miles and 1,440 orbits of the Earth

3 November 1957
Launch of the dog Laika, the first living creature to enter orbit around the Earth. She died within hours, when the cabin of her Sputnik 2 satellite overheated

2 January 1959
Launch of Luna 1, the first spacecraft to depart Earth's gravitational field and reach the distance of the Moon. It is now in heliocentric orbit

12 September 1959
Luna 2 became the first spacecraft to physically impact the Moon, crash-landing in the Mare Imbrium region, close to the craters Aristides, Archimedes and Autolycus

7 October 1959
Never before seen by human eyes, the lunar far side, as seen for the first time by Luna 3, proved to be mountainous, with very few low-lying plains

1960-1969

Jan Feb Mar Apr May June July Aug Sept Oct Nov Dec

12 April 1961
Atop a modified version of Sergei Korolev's R-7 intercontinental ballistic missile, Yuri Gagarin became the first human being to enter space and complete a single Earth orbit

20 February 1962
John Glenn became the first American to orbit the Earth, launching aboard a modified Atlas intercontinental ballistic missile and returning to a splashdown in the Atlantic Ocean

5 May 1961
Three weeks after Gagarin's triumph, Alan Shepard became America's first man in space. He flew a 15-minute suborbital voyage aboard the Freedom 7 capsule

14 December 1962
Mariner 2 became the first spacecraft to successfully encounter another planet when it flew within 18,700 miles of Venus, revealing thick atmospheric clouds

16 June 1963
Former factory worker Valentina Tereshkova was hurriedly trained as part of a propaganda campaign by the Soviet Union to secure a record for the first woman in space

18 March 1965
For 16 minutes, Alexey Leonov floated in the vacuum of space, protected only by his pressurised suit. In doing so, he became the world's first spacewalker

20 July 1969
After millennia of gazing upward at the Moon, the space race officially ended when Neil Armstrong and Buzz Aldrin triumphantly set foot on the Sea of Tranquillity

24 December 1968
On Christmas Eve 1968, Apollo 8 astronauts Frank Borman, Jim Lovell and Bill Anders observed 'Earthrise' from behind the limb of the Moon, for the first time

20 July 1976
Seven years after the first manned Moon landing, Viking 1 became the first spacecraft to soft-land on the surface of Mars and successfully completed its mission

3 December 1973
Pioneer 10 became the first spacecraft to cross the asteroid belt and fly past Jupiter. It revealed the giant planet's punishing radiation belts, which caused several transistors to fail

1970-1979

20 February 1986
Unlike previous 'monolithic' space stations, Russia's Mir complex was intended to be evolvable, with add-on modules. It remained in space for 15 years

19 April 1971
Defeated in the race to the Moon, the Soviets turned their attention to near-Earth projects. They became the first nation to launch a long-duration space station, Salyut 1

1980-1989

1990-1999

14 February 1990
From the very edge of the Solar System, Voyager 1 acquired a 'family portrait', showing six of the then-known planets, minus Mercury, Mars and the dwarf world Pluto

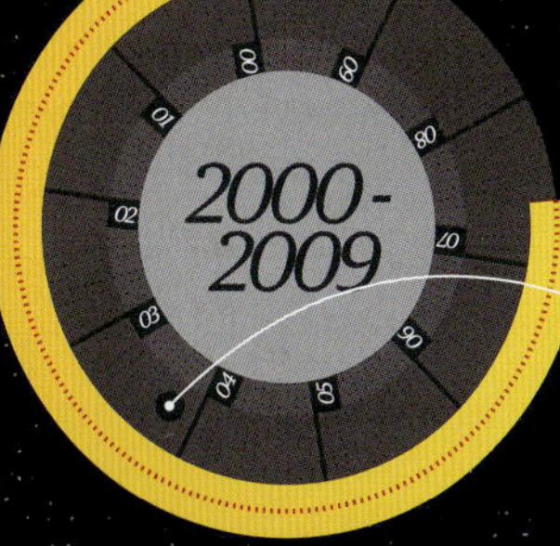

2000-2009

15 July 1965
Mariner 4 became the first spacecraft to successfully observe Mars, returning the first-ever images from deep space. It revealed the planet as cratered and geologically dead

12 April 1981
STS-1, maiden voyage of Columbia, represented the first flight of a reusable winged orbital spacecraft with humans aboard. It marked the dawn of a 135-flight career for the shuttle fleet

15 October 2003
Taikonaut Yang Liwei became the first Chinese spacefarer when China launched the Shenzhou 5 spacecraft, and became the third nation to launch its own personnel into orbit

"TWO TEAMS WORKING SEPARATELY MADE THE DISCOVERY OF THE DNA STRUCTURE IN 1953"

Legend states that Watson and Crick walked into the Eagle pub in Cambridge and proclaimed their discovery to the bar's lunchtime guests

The 1953 discovery wasn't the first study of DNA. In 1869, Friedrich Miescher noticed a new substance while studying proteins

UNLOCKING THE SECRETS OF DNA

"WE HAVE DISCOVERED THE SECRET OF LIFE." HOW THE STUDY OF DNA TURNED THE PAGE ON OUR UNDERSTANDING OF GENETICS AND HOW ALL LIVING THINGS DEVELOP

WRITTEN BY **JACK GRIFFITHS**

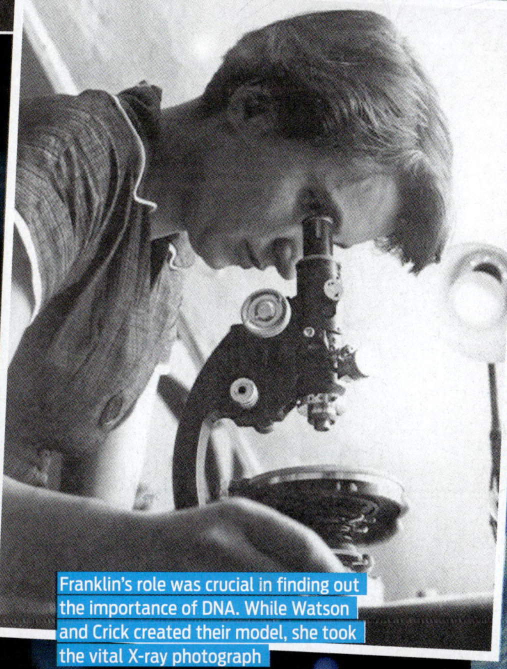

Franklin's role was crucial in finding out the importance of DNA. While Watson and Crick created their model, she took the vital X-ray photograph

The discovery of DNA was a critical moment in human history. It is contained in every cell in the human body, is passed down generations and carries the genetic information which drives evolution. We share most our DNA with chimpanzees and even some with the humble banana.

DNA was first found in 1869 when Swiss physician Friedrich Miescher was studying proteins in cells and noticed a new substance. Calling it nuclein and then nucleic acid, he noted its finding and how to extract it, but not its purpose. It was not until the next century that it was named as deoxyribonucleic acid or DNA and its function was understood. Further progress came in 1943 when British scientist William Astbury came close to identifying the structure of DNA using X-Ray diffraction. A year later, three US/Canadian scientists agreed that DNA contained genetic information.

Two teams working separately in Cambridge and London made the discovery of the DNA double helix structure in 1953. James Watson and Francis Crick were creating models in the university city to illustrate the shape of DNA while Maurice Wilkins and Rosalind Franklin were in the capital's King's College studying it using X-ray crystallography. Franklin saw the famous DNA helix shape through a X-ray photograph that she took. This was sent to Watson and Crick by Wilkins who used this to prove that their models were correct. The three men scooped the 1962 Nobel Prize for their work but Franklin was not honoured. She died of cancer five years later aged 37 and never received the accolade she deserved.

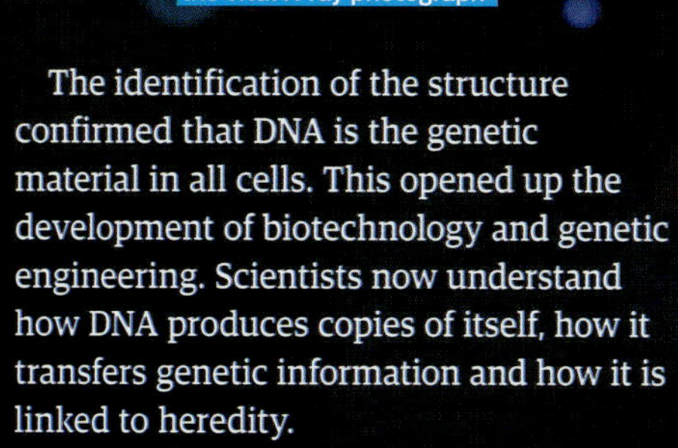

The identification of the structure confirmed that DNA is the genetic material in all cells. This opened up the development of biotechnology and genetic engineering. Scientists now understand how DNA produces copies of itself, how it transfers genetic information and how it is linked to heredity.

The knowledge of DNA allows us to identify evidence for criminal cases and to establish the parentage of children. Gene therapy can help prevent diseases and mutations such as some forms of cancer and sickle cell disease. This is achieved by replacing defective or missing genes with newer, healthier ones to help boost the body's defences. We also have the ability to alter a cell's function which is very useful in creating pest-resistant crops, more efficient and healthier livestock and improved performance of biofuels.

WORLDS WITHIN WORLDS

DRILLING DOWN TO THE FUNDAMENTAL BUILDING BLOCKS OF THE UNIVERSE

WRITTEN BY **EDOARDO ALBERT**

Back in the first half of the 20th century, it seemed like the stuff that made up the universe was pretty simple. Everything was made of atoms, and each atom was made of a core of protons and neutrons, surrounded by a whirling orbit of electrons: a set-up sort of like the Solar System in miniature.

Only it turned out it wasn't so simple after all. When scientists started smashing particles into each other in high-energy accelerators, they started discovering a whole range of particles that they hadn't expected. By the 1960s, the number was becoming dizzying.

This was a long way from the elegant and simple solution of protons, neutrons and electrons and, for the physicists of the time, it seemed clear that they were staring at trees and missing the forest.

Through a worldwide scientific effort, combining theory and experimentation, a new model of the building blocks of the universe was devised. This Standard Model (the name was first used by Steven Weinberg in 1973 and popularised in

STANDARD MODEL

The Standard Model. Six quarks and six leptons, which together make up the fermions, and four bosons

This diagram graphically conveys the elegance of the Standard Model

1975 by Sam Treiman and Abraham Pais) simplified the stew of particles to just 17 fundamental particles. Almost all of these were new. Only the electron and the photon were carried over from older physics models.

The 17 fundamental particles are split into two groups: the fermions and the bosons. The 12 fermions consist of six quarks and six leptons. Various combinations of these produce the standard protons and neutrons familiar from older theories and together they make up the mass of the visible universe.

The five bosons, on the other hand, are the workers. They mediate three of the four fundamental forces of the universe: the strong nuclear force, the weak nuclear force and electromagnetism.

The strong nuclear force is the one that holds the nucleus of an atom together against the electromagnetic repulsion of similarly positively charged particles and it's carried by the well-named gluon.

The weak nuclear force produces radioactive decay and it's mediated by W and Z bosons, while old-school photons carry the electromagnetic force. The final boson is the Higgs boson, which provoked a long search for evidence of its existence.

The Standard Model has been incredibly successful in describing the physics of the subatomic realm. 115 years of work went into it, from the discovery of the electron in 1897 to confirmation of the Higgs boson in 2012.

However, the Standard Model doesn't explain everything. The fourth fundamental force, gravity, does not fit into it at all. What's worse, we've recently discovered that we've only been looking at a tiny fraction of everything. Planets, stars, galaxies, all the stuff that astronomers and physicists have dealt with throughout history comprises only five per cent of the mass of the universe. There's a further 27 per cent made of dark matter and an extraordinary 68 per cent produced by dark energy. That's 95 per cent of the universe that we've only just discovered! So it's safe to say we're a little way away from a theory of everything.

Images source: Adobe Stock, Alamy

THE THEORIES OF STEPHEN HAWKING

SOME OF THE LATE PHYSICIST'S THEORIES REVOLUTIONISED THE WAY WE VIEW THE UNIVERSE, BUT OTHERS STILL LEAVE SCIENTISTS SCRATCHING THEIR HEADS

WRITTEN BY **ANDREW MAY**

HAWKING'S CV

1965-1969

Research fellow, Gonville and Caius College, Cambridge

Extended the concepts of the singularity theorem explored in his PhD thesis and explored the idea that the universe may have begun as a singularity.

1969-1975

Fellowship for distinction in science, Gonville and Caius College, Cambridge

A post created specially for him. Here he proposed the laws of black hole mechanics and discovered Hawking radiation.

1975-1979

Reader in gravitational physics and professor of gravitational physics, Gonville and Caius College, Cambridge

Published his first book. Won Eddington, Maxwell, Hughes, Albert Einstein and Pius XI medals, plus the Dannie Heineman Prize. Received an honorary doctorate from the University of Oxford.

1979-2009

Lucasian professor of mathematics at the University of Cambridge

Pushed the boundaries of knowledge of black holes and the beginning of the universe. Published many books, including a 10-million-copy bestseller.

2009-2018

Director of research, Cambridge University Department of Applied Mathematics and Theoretical Physics

Published more books, inspired a Hollywood movie, held a party for time travellers; received the Presidential Medal of Freedom, Russian Special Fundamental Physics Prize and an honorary doctorate from Imperial College London.

TIMELINE

OCTOBER 1959

Began studying at University College, Oxford, aged 17.

OCTOBER 1962

Became a doctoral student at Cambridge under Dennis Sciama.

FEBRUARY 1963

Diagnosed with amyotrophic lateral sclerosis following a fall while ice skating.

"HIS THEORIES SEEMED BIZARRELY FAR OUT AT THE TIME HE FORMULATED THEM"

PROVEN

THE BIG BANG

Hawking got off to a flying start with his doctoral thesis, written at a critical time when there was a heated debate between two rival cosmological theories: the Big Bang and the steady-state universe. Both theories accepted that the universe is expanding, but in the first it expands from an ultra-compact, super-dense state at a finite time in the past, while the second assumes the universe has been expanding forever, with new matter always being created to maintain a constant density. In his thesis, Hawking showed that steady-state theory is mathematically self-contradictory. He argued instead that the universe began as an infinitely small, infinitely dense point called a singularity. Today Hawking's description is almost universally accepted among scientists.

PROVEN

BLACK HOLES ARE REAL

More than anything else, Hawking's name is associated with black holes - another kind of singularity, formed when a huge star undergoes complete collapse under its own gravity. These mathematical curiosities arose from Einstein's theory of general relativity, and they had been debated for decades when Hawking turned his attention to them in the early 1970s.

His stroke of genius was to combine Einstein's equations with those of quantum mechanics, turning what had previously been a theoretical abstraction into something that looked like it might actually exist in the universe. The final proof that Hawking was correct came in 2019, when the Event Horizon Telescope obtained a direct image of the supermassive black hole lurking in the centre of giant galaxy Messier 87.

PROVEN

HAWKING RADIATION

Black holes got their name because their gravity is so strong that photons - or particles of light - shouldn't be able to escape from them, rendering them dark. But in his early work on the subject, Hawking argued that the truth is more subtle than this monochrome picture.

By applying quantum theory - specifically the idea that pairs of 'virtual photons' can spontaneously be created out of nothing - he realised that some of these photons would appear to be radiated from the black hole. Now referred to as Hawking radiation, the theory was recently confirmed in a laboratory experiment at the Technion - Israel Institute of Technology in Haifa. In place of a real black hole, the researchers used an acoustic analogue: a 'sonic black hole' from which sound waves cannot escape. They detected the equivalent of Hawking radiation exactly in accordance with the physicist's predictions.

"BY APPLYING QUANTUM THEORY, HE REALISED THAT SOME OF THESE PHOTONS WOULD APPEAR TO BE RADIATED FROM THE BLACK HOLE"

PROVEN

BLACK HOLE AREA THEOREM

In classical physics, entropy, or the disorder of a system that can only ever increase with time, never decreases. Together with Jacob Bekenstein, Hawking proposed that the entropy of a black hole is measured by the surface area of its surrounding event horizon.

The recent discovery of gravitational waves emitted by merging pairs of black holes shows that Hawking was right again. As Hawking explained to the BBC after the first such event in 2016: "The observed properties of the system are consistent with predictions about black holes that I made in 1970. The area of the final black hole is greater than the sum of the areas of the initial black holes." More recent observations have provided further confirmation of Hawking's area theorem.

1970
With Penrose, he published a proof that the universe must have begun as a singularity.

1973
He published his first book, *The Large Scale Structure of Space-Time*, cowritten with George Ellis.

1974
He was elected a fellow of the Royal Society just weeks after the announcement of Hawking radiation.

1979
Elected Lucasian professor of mathematics at the University of Cambridge.

1981

Proposed that information in a black hole is irretrievably lost when a black hole evaporates, igniting a 'black hole war' with Susskind and 't Hooft.

1984

First draft of *A Brief History of Time* completed, but the publisher thinks it's too technical.

1985

Contracted pneumonia and underwent a tracheotomy, losing his power of speech.

1988

A Brief History of Time was published. The book went on to sell 10 million copies worldwide.

1988

Together with Penrose, he won the Wolf Prize for physics, worth $100,000, "for their brilliant development of the theory of general relativity".

1989

Appointed a Companion of Honour in the Birthday Honours, but turned down a knighthood.

UNPROVEN

DOOMSDAY PROPHECIES

In his later years, Hawking made a series of bleak prophecies concerning the future of humanity that he may or may not have been totally serious about. These range from the suggestion that the elusive Higgs boson, or 'God particle', might trigger a vacuum bubble that would gobble up the universe, to hostile alien invasions and artificial intelligence (AI) takeovers. Although Hawking was right about so many things, we'll just have to hope he was wrong about these.

PROVEN

THE INFORMATION PARADOX

The existence of Hawking radiation creates a serious problem for theoreticians. It seems to be the only process in physics that deletes information from the universe. The basic properties of the material that went into making the black hole appear to be lost forever - the radiation that comes out tells us nothing about them.

This is the so-called information paradox that scientists have been trying to solve for decades. Hawking's own take on the mystery, which was published in 2016, is that the information isn't truly lost. It's stored in a cloud of zero-energy particles surrounding the black hole, which he dubbed 'soft hair'. But Hawking's hairy black hole theorem is only one of several hypotheses that have been put forward, and to date no one knows the true answer.

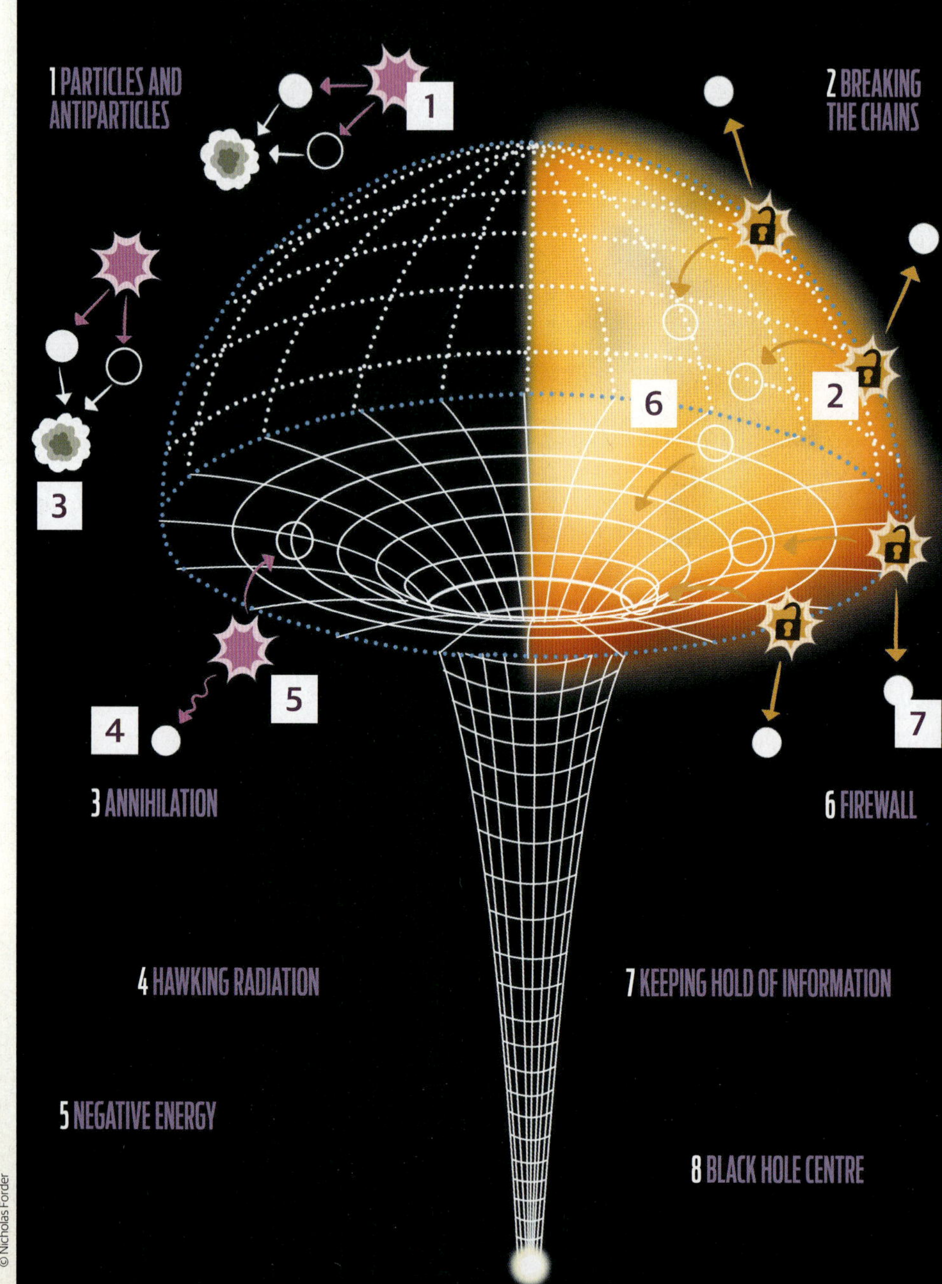

© Tobias Roetsch

UNPROVEN

THE EXISTENCE OF A MULTIVERSE

One of the topics Hawking tinkered with towards the end of his life was multiverse theory - the idea that our universe, with its beginning in the Big Bang, is just one of an infinite number of coexisting bubble universes. Hawking wasn't happy with the suggestion made by some scientists that any ludicrous situation you can imagine must be happening right now somewhere in that infinite ensemble.

In his very last paper in 2018, Hawking sought, in his own words, to "try to tame the multiverse". He proposed a novel mathematical framework that, while not dispensing with the multiverse altogether, rendered it finite rather than infinite. But as with any speculation concerning parallel universes, we have no idea if his ideas are right. And it seems unlikely that scientists will be able to test his idea any time soon.

"HAWKING SOUGHT TO TRY TO TAME THE MULTIVERSE"

1991
A film version of *A Brief History of Time* premiered.

1997
A six-part documentary, *Stephen Hawking's Universe*, was made.

2004
Admitted he was on the wrong side in the 'black hole war', and bought Preskill an encyclopedia.

2005
Published *A Briefer History of Time* with Leonard Mlodinow to update his ideas and make them more accessible.

2007
Published *George's Secret Key to the Universe* with his daughter.

2009
Received the Presidential Medal of Freedom from Barack Obama.

2009
Held a party for time travellers, but none showed up.

2018
Hawking passed away.

UNPROVEN

PRIMORDIAL BLACK HOLES

Hawking was the first to explore the theory behind primordial black holes in depth. It turns out they could have any mass - though really tiny ones would have 'evaporated' into nothing by now due to Hawking radiation. One possibility considered by Hawking is that primordial black holes might make up the mysterious dark matter that astronomers believe permeates the universe. But current observational evidence indicates this is unlikely. Either way, we currently don't have observational tools to detect primordial black holes.

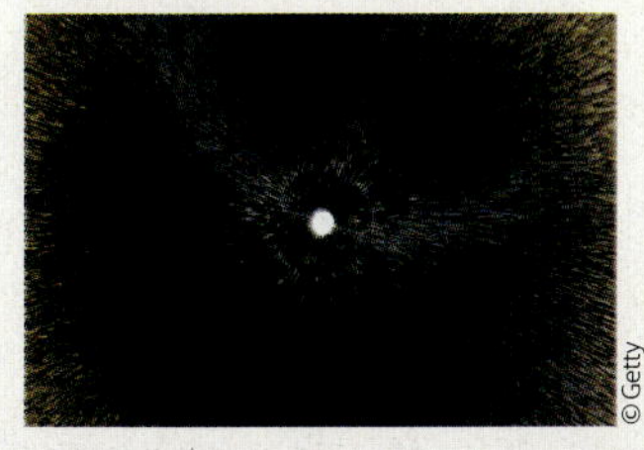

© Getty

UNPROVEN

IS TIME TRAVEL POSSIBLE?

Surprising as it may sound, the laws of physics as we understand them today don't prohibit time travel. The solutions to Einstein's equations of general relativity include 'closed timelike curves', which would effectively allow you to travel back into your own past. Hawking was bothered by this because he felt that backward time travel raised logical paradoxes that shouldn't be possible. He suggested that some currently unknown law of physics prevents closed timelike curves from occurring - his so-called 'chronology protection conjecture'.

© Getty

UNPROVEN

IS THERE A CREATOR?

One question cosmologists get asked most often is about what happened before the Big Bang. Hawking's own view was that the question is meaningless. For all intents and purposes, time itself - as well as the universe and everything in it - began at the Big Bang. "For me, this means that there is no possibility of a creator," he wrote in his final book, "because there is no time for a creator to have existed in." That's an opinion many people will disagree with, but one that Hawking expressed on numerous occasions throughout his life.

© Getty

Dolly's white face proved that she was a clone, and not a genetic relation to her mother, who had a black face

DOLLY THE SHEEP

THE STORY OF THE FAMOUS SCOTTISH EWE THAT HAD A HUGE IMPACT ON THE PROGRESS OF CLONING TECHNIQUES AND BOOSTING THE KNOWLEDGE OF GENETIC ENGINEERING

WRITTEN BY **JACK GRIFFITHS**

She is the world's most famous sheep and her creation by the University of Edinburgh's Roslin Institute, helped to progress genetic engineering

Dolly was the first adult mammal cloned from an adult cell. A tadpole, carp, mouse and a cow predate her via other cloning methods

Genetic engineering took a giant leap forward in the summer of 1996. As part of a study to create genetically-modified livestock, Dolly the sheep became the first mammal successfully cloned from an adult cell.

Dolly was cloned from the cells of two female sheep using somatic cell nuclear transfer (SCNT). The first supplied an udder cell, while the other provided an unfertilised egg. After the nucleus was detached from the egg, the two were electrically-fused together so it now had Dolly's DNA. This cell then started dividing and these were placed into 13 other sheep. One of these ewes became pregnant and, five months later on 5 July, Dolly was born.

Named after the American country singer Dolly Parton, the birth was revealed to the world in February 1997. The media descended on Roslin, Scotland where the institute was located, to meet the famous sheep. The next stage of the process was for Dolly to procreate, and she duly did the following year, eventually having six lambs with a ram called David.

The process raised questions over the ethics of cloning and how it has the potential to be misused. The success demonstrated that certain traits could be transmitted to other animals and it was also discovered that animal cells had more information contained in them than first thought. It was believed that each cell, for example blood or pancreatic, could only undertake one function. However, Dolly proved that just one mammary gland could create a whole cloned animal.

Since Dolly, other sheep, as well as larger mammals like cattle, wolves and horses have been successfully cloned through SCNT, and there have also been duplicates made of human stem cells. Farming has reaped the benefits, cloning livestock with the best genes for food production or resisting disease. Dolly has paved the way for theories on how to clone endangered species like the white rhino and even extinct and prehistoric animals.

Sheep have an average natural life expectancy of 10-12 years but Dolly died aged six in 2003. She had been ill since late 2001 when she developed arthritis but later began coughing regularly and a scan revealed that she had incurable tumours in her chest. Dolly was never awoken from the general anaesthetic she had for the scan and, after death, was donated to the National Museum of Scotland in Edinburgh where she is still on display today.

Images source: Adobe Stock, Alamy

THE HIGGS MECHANISM

THE REALLY, REALLY COMPLICATED PART OF THE STANDARD MODEL OF PARTICLE PHYSICS

WRITTEN BY **EDOARDO ALBERT**

Yes, this is all about the God particle. It was named that in a 1993 book by Leon Lederman, who won the Noble Prize for physics in 1988. The man after whom the particle is otherwise named, Peter Higgs, was not too pleased with the new name but it was a godsend (sorry!) for journalists looking for snappy headlines, so it has stuck.

Another, less often given, reason for the name is that it takes almost superhuman levels of intelligence to actually understand what the Higgs boson and its associated Higgs field are and what they do. Technically, it involves quantum field theory and the symmetry of massless particles. Both of these are formidably difficult.

So a better way to understand the Higgs field is to look at what the physicists who first proposed the idea in 1964 were trying to do. There were three separate groups of researchers who all came up with versions of the idea: Robert Brout and François Englert; Gerald Guralnik, Carl Richard Hagen and Tom Kibble; and then Peter Higgs working all by himself.

These scientists were trying to explain why the weak nuclear force acts only at very short range. Gravity and electromagnetism work at long ranges, their force gradually declining according to the inverse square law. But both the strong and weak nuclear forces operate at very very short distances, basically the diameter of the nucleus of an atom. Why was this so?

This is what they proposed, turned into physical analogues we can understand. Think of a lamp, shining in the middle of a field. At night you can see it clearly but it gradually gets dimmer as you walk away: this is like gravity and electromagnetism. Strong up close and then gradually fading.

But suppose you lit a fire under the lamp so that smoke started rising up past it. If the smoke was thick enough, it would quickly obscure the lamp because the smoke absorbs the light. This is the Higgs mechanism. It's a field which acts like the smoke around the lamp, absorbing the weak nuclear force.

To prove this theory, however, physicists had to detect the Higgs boson. This was difficult because the Higgs boson is the ultimate flyby guest: it's barely there before it's gone, decaying into other particles. Indeed, it decays so quickly that it can't be detected directly: its existence has to be inferred from what it leaves behind, its decay products. After decades of work, the Higgs boson, or rather, its shadow, was detected at CERN in Geneva in 2012.

The God particle was real. And, therefore, the Higgs mechanism was working as Higgs and co had predicted. Score up another point to the Standard Model of particle physics.

Peter Higgs and the Large Hadron Collider at CERN, Geneva

"AFTER DECADES OF WORK, THE HIGGS BOSON, OR RATHER, ITS SHADOW, WAS DETECTED"

Images Source: Adobe Stock, Getty

Future PLC Quay House, The Ambury, Bath, BA1 1UA

Editorial
Editor **April Madden**
Senior Designer **Perry Wardell-Wicks**
Head of Art & Design **Greg Whitaker**
Editorial Director **Jon White**
Managing Director **Grainne McKenna**

Contributors
Edoardo Albert, Marc DeSantis, Ben Evans, Ben Gazur, Jack Griffiths, Andrew May, Laura Mears, Poppy-Jay Palmer, Jodie Tyley and Derek Wilson

Cover images
Adobe Stock, Getty images

Photography
All copyrights and trademarks are recognised and respected

Advertising
Media packs are available on request
Commercial Director **Clare Dove**

International
Head of Print Licensing **Rachel Shaw**
licensing@futurenet.com
www.futurecontenthub.com

Circulation
Head of Newstrade **Tim Mathers**

Production
Head of Production **Mark Constance**
Production Project Manager **Matthew Eglinton**
Advertising Production Manager **Joanne Crosby**
Digital Editions Controller **Jason Hudson**
Production Managers **Keely Miller, Nola Cokely, Vivienne Calvert, Fran Twentyman**

Printed in the UK

Distributed by Marketforce – www.marketforce.co.uk
For enquiries, please email: mfcommunications@futurenet.com

History of Science First Edition (AHB6377)

Future plc is a public company quoted on the London Stock Exchange (symbol: FUTR)
www.futureplc.com

Chief Executive Officer **Jon Steinberg**
Non-Executive Chairman **Richard Huntingford**
Chief Financial Officer **Sharjeel Suleman**

Tel +44 (0)1225 442 244